Business Pro___ __na

A Rigorou_ Ap

The British Computer Society

The British Computer Society is the leading professional body for the IT industry. With members in over 100 countries, the BCS is the professional and learned Society in the field of computers and information systems.

The BCS is responsible for setting standards for the IT profession. It is also leading the change in public perception and appreciation of the economic and social importance of professionally managed IT projects and programmes. In this capacity, the Society advises, informs and persuades industry and government on successful IT implementation.

IT is affecting every part of our lives and that is why the BCS is determined to promote IT as *the* profession of the 21st century.

Joining the BCS

BCS qualifications, products and services are designed with your career plans in mind. We not only provide essential recognition through professional qualifications but also offer many other useful benefits to our members at every level.

Membership of the BCS demonstrates your commitment to professional development. It helps to set you apart from other IT practitioners and provides industry recognition of your skills and experience. Employers and customers increasingly require proof of professional qualifications and competence. Professional membership confirms your competence and integrity and sets an independent standard that people can trust. www.bcs.org/membership

Further Information

Further information about the British Computer Society can be obtained from: The British Computer Society, 1 Sanford Street, Swindon, Wiltshire, SN1 1HJ, UK.
Telephone: + 44 (0)1793 417 424
Email: bcs@hq.bcs.org.uk
Web: www.bcs.org

Business Process Management

A Rigorous Approach

Martyn A Ould
Venice Consulting Ltd

THE BRITISH COMPUTER SOCIETY

First published in the UK by:
The British Computer Society,
1 Sanford Street,
Swindon, Wiltshire SN1 1HJ,
UK
www.bcs.org

ISBN 1-902505-60-3

Co-published in the USA and Canada by:
Meghan - Kiffer Press
310 Fern Street
Tampa, FL 33609,
USA
www.mkpress.com

ISBN (USA and Canada) 0-929652-27-4

British Cataloguing in Publication Data.
A CIP catalogue record for this book is available at the British Library.

All trademarks, registered names etc acknowledged in this publication are to be the property of their respective owners.

Disclaimer:
Although every care has been taken by the authors and The British Computer Society in the preparation of the publication, no warranty is given by the authors or The British Computer Society as Publisher as to the accuracy or completeness of the information contained within it and neither the authors nor The British Computer Society shall be responsible or liable for any loss or damage whatsoever arising by virtue of such information or any instructions or advice contained within this publication or by any of the aforementioned.

Typeset by Tradespools, Frome, Somerset.
UK edition:
Printed at Antony Rowe Ltd., Chippenham, Wiltshire.

Contents

Figures

Author

Martyn Ould read mathematics at Cambridge University and entered the software industry directly in 1970, working for several years for ICL on operating systems. After a short spell at King's College Hospital Computer Centre, he worked for eleven years with Logica, principally on real-time systems, as developer, designer and project manager. In his last two years there he co-founded a company-wide software engineering initiative and co-wrote his first book *A Practical Handbook for Software Development* (Ould and Birrell, 1985 and 1988). His second book, *Testing in Software Development* (Ould and Unwin, 1986 and 1988) resulted from work done by the BCS Testing Working Group.

In 1985 he joined software house Praxis where he became Quality and Technical Director, with responsibility for the company's quality policy and strategy and its overall technical strategy. He developed a systematic planning technique – Strada – for software engineering projects described in his fifth book *Managing Software Quality and Business Risk* (Ould, 1999). At Praxis he consulted for clients on the software engineering process, specialising in project audits and rescues, reviews of software development methods, and risk management. He continues this work as an independent consultant, and teaches Strada on graduate courses at Oxford University.

At Praxis he led the development of the STRIM method for business process design and diagnosis described in his fourth book *Business Processes* (Ould, 1995). He has since extended this to the full-spectrum business process management method – Riva – described in this book. Following the merger with the Deloitte & Touche Consulting Group, he was a Senior Manager within the global firm. He became an independent consultant in 1998 and now consults in business process management and provides training in Riva. He has contributed numerous articles, reviews and papers to the computing press; lectured to public, government, university and corporate audiences; and acted as a referee for national research programmes, conferences and journals.

He is a Fellow of the BCS and a Chartered Engineer.

In his spare time he is a letterpress printer, designing, printing and publishing limited edition books with commissioned illustrations using traditional metal type and mechanical presses. Details of his consulting services and printing and publishing activities can be found at www.ould.org.

Abbreviations

BPEL	Business Process Execution Language
BPM	Business Process Management
BPML	Business Process Modeling Language
BPMN	Business Process Modeling Notation
BPMS	Business Process Management System
BPR	Business Process Re-engineering
CMP	Case Management Process
CP	Case Process
CPA	Critical Path Analysis
CRF	Case Report Form
CRO	Clinical Research Organization
CSP	Case Strategy Process
DBMS	Database Management System
EBE	Essential Business Entity
ELH	Entity Life History
ER	Entity Relationship
ERP	Enterprise Resource Planning
IS	Information System
PAD	Process Architecture Diagram
PDA	Personal Digital Assistant (handheld device)
QMS	Quality Management System
RAD	Role Activity Diagram
RML	Requirements Modeling Language
SAS	System Attribute Specification
SOP	Standard Operating Procedure
SPML	STRIM Process Modelling Language
SSADM	Structured Systems Analysis and Design Methodology
STRIM	Structured Technique for Role Interaction Modelling (the precursor of Riva)
TQM	Total Quality Management
UML	Unified Modeling Language
UOW	Unit of Work
WFM	Workflow Management

Preface

Much has happened in the business process world since I wrote *Business Processes* (Ould, 1995). The recent growth of interest in the important emerging technology of Business Process Management and the experience I have had using the approach described in this book have together conspired to force me back to my desk to write this current work about the Riva method.

My grateful thanks go in particular to Stewart Green and his colleagues (University of the West of England, Bristol), Tim Huckvale (Charteris), Dr Steve Hutson (Calcis Consulting Group), Derek Miers (Enix Consulting Limited and Director of BPMI.org), David Perrin (Holonix Limited), Clive Roberts (Co-ordination Systems Limited), Howard Smith (Co-chair of BPMI.org and CTO Computer Sciences Corporation European Group) and Professor Bob Snowdon (University of Manchester). Michael Jackson's work on system development has been very influential. Their help on the text, and conversations with them and many others over the years have helped me where private cogitation has failed. Many of their thoughts, ideas and words have found their way into the melting pot and it is hard now to identify, let alone acknowledge, all of them. Insights arise from many sources, not least from application to real problems but also from the courses I have given on Riva and the challenges of course attendees. Any nonsense is, of course, mine.

I would also like to thank a number of people for permission to use material: Deloitte for allowing me to carry forward some material from my 1995 book; Howard Smith for quotations from his and Peter Fingar's influential book *Business Process Management: The Third Wave* (Smith and Fingar, 2002) and other articles; Tony Solomonides at the University of the West of England, Bristol, for permission to use the Sentinel case study in Chapter 7; the University of the West of England, Bristol for permission to use case study material from work undertaken with the CEMS Faculty.

My website at www.veniceconsulting.co.uk contains free downloads of some resources to support the use of Riva. Questions, criticisms and contributions are welcome and should be addressed to riva@venicecon-sulting.co.uk.

Martyn Ould
Hinton Charterhouse, 1 May 2004

Introduction

PROCESSES ARE BACK ON TOP

Something very different is happening in the world of process. In recent decades business processes have ridden two 'waves'. A third is upon us.

In the first wave, processes existed purely in terms of custom – 'it's just the way we do things round here' – or, if they were lucky, lived a private life inside policy and procedure manuals. The emphasis was towards 'small' pieces of organizational activity such as claiming expenses and approving budgets, or perhaps standardized and often-repeated sequential activity such as the analysis of a laboratory sample. If processes were modelled it was with simple diagrams such as flowcharts. And perhaps flowcharts were enough for simple sequential processes. The pictures were stored in procedure manuals – and all too easily ignored. Sometimes those pictures would be drawn as part of an attempt to understand the process and how it might be improved, how things might be done more quickly or more cheaply: in other words for process improvement. Total Quality Management (TQM), with its Seven Tools of Quality, used simple models to examine processes, looking for those – typically small-scale – improvements.

In the second wave, processes initially became the unwitting victims of the information system developers: 'information' was on top. By cementing the allowable information flows in the information system, the information system engineers cemented the business process at the same time. The process could no longer be changed without expensive re-engineering of the underlying information systems. But worse was to come: processes found themselves captives of Enterprise Resource Planning (ERP) systems. A side effect of Business Process Re-engineering (BPR) was that many of the 'common' processes of organizations were re-engineered to what was (not jokingly) called 'best practice' or 'world-class practice'. Recognizing that the commonality of such processes meant that they could be made into commodities, vendors of ERP systems defined processes for whole areas of organizational activity and built them into their systems – everyone could be offered the same 'competitive advantage'! The 'Inventory Management Process' could be bought by all and sundry. That's '*The* Inventory Management Process'. Once again, the process had become a victim, condemned to second-class citizenship, subservient to 'information'.

But we are now seeing a Third Wave – a term first coined by Howard Smith and Peter Fingar (Smith and Fingar, 2002) – a wave driven by the new technology of Business Process Management (BPM). Finally, the business process has become a first-class citizen, the one that now determines what information will be kept: process first, then information. Instead of being buried in the rules of a relational database or in the settings of an ERP system, invisible but cast in concrete, the process is now visible, changeable, and potentially back in the hands of the organization, with the result that the organization now has the power to change how it wants to do its business, and to change it when it needs to.

At the heart of BPM is a different understanding of business processes. Part of that understanding is that our process is not something that could perhaps be 'deduced' from the way our information system is set up or from what our ERP allows us to do. It is not 'implied by' the information system. Our process has its own separate existence in a form that – given to a 'process enactment engine' – can be executed or 'run', that can be changed on-the-fly, that can be evolved as our business evolves, that can be monitored in real time and that can be deployed at will through the organization. A computer system that supports our organization no longer simply helps us to manage our information: it now helps us, first and foremost, to manage our processes. It is a Business Process Management System (BPMS).

This third wave needs appropriate methods for thinking about processes, for working with processes, for defining, designing and analyzing processes in a way that positions us to use those new BPM systems. If you are familiar with my previous book *Business Processes* (Ould, 1995) you will know that such a method has been around for some time, originally in the form of a Structured Technique for Role Interaction Modelling (STRIM) and now in its updated form Riva. This book describes Riva.

Riva has its roots in the IPSE 2.5 research project carried out as part of the UK Alvey Programme in 1986. As part of the project, Clive Roberts and I undertook to develop a language that, firstly, could be used to describe a process and, secondly, was defined to a point where a process model written in that language could be given to a computer system which would then 'enact' the process, thereby supporting the group who would collectively carry it out. The solution we developed was a combination of Anatol Holt's Role Activity Diagrams (RADs), to which we made some important adaptations, and Sol Greenspan's Requirements Modeling Language (RML), from which we developed the STRIM Process Modeling Language (SPML), in particular by adding the concept of *role* to its formal semantics. Central to IPSE 2.5 was the idea that a process model could be changed on-the-fly by the process users and while the process was running. (References to relevant literature can be found at the end of this chapter.) IPSE 2.5 was an early BPMS, but it has taken the intervening decades to see the development of technologies necessary to make it a reality, not least the internet, web services, and much of the object-oriented application infrastructure now available.

When Clive and I developed SPML we had to provide formal semantics for the language so that it was executable and supported on-the-fly changes. We also defined transformation rules that allowed us to translate a RAD into SPML; this meant that a RAD had strong semantics itself, and it is this feature that is important in the area of modelling to enact a process. SPML has now been taken forward and the new version is called CoSpeak – it is a language of coordination. To the everyday user of Riva, the presence of an XML variant called CoSpeak is not important. But to the analyst and to those interested in building BPMSs, the presence of a formal language with full semantics, underpinning RADs, is vital. Without those semantics we cannot say unambiguously what a given RAD means in terms of the behaviour of the process it describes.

Holt himself describes the application of his Role/Activity Theory to *coordination systems* in terms of a conceptual model based around what people as a group *do*, rather than around the data they operate on. It is hard to overemphasize this point. A lot of process modelling has come from the software engineering world where, historically, data and information have ruled. That sort of process modelling has therefore concentrated on things and data about things. But processes are about dynamics, activity, collaboration and cooperation. So the way we think about processes must have these at the centre. We must put processes back on top. Riva does this.

WHAT A 'PROCESS' IS AND WHAT IT ISN'T

I'm a stickler for good definitions. I believe that we can get a long way in solving many problems by defining our basic concepts properly. By striving for clarity in those basic concepts we force ourselves to deepen our understanding of our subject. Moreover, those definitions are the foundations of our work: poor definitions mean poor theories, and poor theories make poor practice. I don't want to assemble a lot of shoddy ideas and call them a method. I want to build a *principled* theory and practice of business processes. The more principled our theory is, the more powerful it will be and the more quickly we can get to results. I also want it to be as simple as possible, but not too simple, as someone cleverer than I once said.

This leaves me wanting to start this book by defining the concept 'process'. If we don't know – from the outset – what the word means, then we won't get very far.

Every organization does things in order to achieve its objectives. For instance:

- We handle orders for goods.
- We recruit staff.
- We design new products.
- We run an investment portfolio.
- We develop new pharmaceutical drugs.

Something that characterizes all of these is they are 'quite large things'. They probably involve more than one person, take more than a moment, and might be carried out in different ways in different situations. We can appreciate that they are different from something like 'Fill in expenses form'. When we use the word 'process' we are thinking of *a coherent body of organizational activity*: activity that goes on in the organization and that in some sense comes as a unit. Typically 'comes as a unit' means 'is all focused on a certain outcome'. For instance in the above cases the outcomes might be:

- To respond to a customer order by shipping the requested goods and invoicing the customer for payment.
- To respond to the staffing needs of the organization by engaging staff of the right type and capabilities on appropriate terms and conditions.
- To answer a gap in the market place with a product that can be manufactured, marketed and sold profitably.
- To decide how available funds will be allocated to financial instruments in order to realize gains of the right value at an acceptable level of risk.
- To develop and bring to market new pharmaceutical drugs that are efficacious and safe.

Let's look at some of the essential features of a process. A process involves *activity*: people and/or machines *do* things. A process also generally involves more than one person or machine working together: a process is about *groups*; in particular it is about *collaborative* activity. And a process has a *goal*: it is intended to achieve something. The group collaborate to achieve the goal.

We can also characterize the concept 'process' by what it is not. It is not the same as a 'functional group', e.g. Personnel, Manufacturing, Finance, Goods Inwards, or Credit Control. These are parts of the organization which have responsibilities, staff and resources; but they are not processes though they take part in processes. In fact, we shall see later that the relationship between processes and functional groups can be complex and indeed that the efficient operation of processes can be hindered by an organization's structures. In a re-engineering context we shall want to explore that relationship between the organization and the process. So, we shall resist all thoughts of the 'Finance process' or the 'HR process' – these are meaningless phrases.

This discussion begs a question: 'How do we sensibly chunk all of the organizational activity into things we can call processes?' This is a hard question, and one that is often badly – even wrongly – answered in business process management projects. Key to it, I believe, is that the method we use must give the answer appropriate to the business of the particular organization we are looking at. And if *you* use that method and *I* use that same method, we should surely both end up with the same answer. If we reach different answers, what value can we give to either?

Suppose I walk into the dissecting room at the teaching hospital to lecture to medical students on how the human body is constructed. Awaiting my arrival is my assistant, ready with a thick marker pen. There, on the table, is the cadaver. I have brought with me an axe. With a deft overhead blow I lop off the lower part of one leg. My assistant labels it 'The A bit'. The students dutifully record the name against their sketches of the body. Some aren't sure quite where on the leg the axe fell, but choose a point anyway. With no more ado, another blow removes the top of the skull, which my assistant labels 'The B bit'. More scratching in notebooks. Further blows yield bits C, D, and E, the remainder on the table being labelled 'The F bit' with the marker pen. 'The F bit' is still quite large, so four swipes render it into five pieces, which my assistant labels 'The F1 bit', 'The F2 bit', and so on. The students take notes, increasingly unsure about exactly how much corpse each bit is made up of. Never mind, I now take bit F2, and putting down the axe in favour of a small meat cleaver, I cleave it into bits whose names, I tell the students, are F2a, F2b, and F2c.

What understanding do the students now have about the way the body is constructed and how it works? Has the chunking been guided by an understanding of what a human body is all about? Would each student have the same understanding of exactly what constituted each bit? If I gave the same lecture next week, would the students' drawings be anything like those of this week's?

You would, I know, prefer that I had taken a scalpel with me to the dissecting room, together with an understanding of what a human body is all about – the fact that there are 'natural cleavage lines' that separate the central nervous system, the gastro-intestinal system, the skeleton, the musculature – and of how those systems are connected. We look for these things because we know that a human body is 'in the business of' feeling and sensing, nourishing itself, standing and moving.

When we chunk organizational activity we shall need a similar scalpel that will allow us to cut along the natural cleavage lines of that activity, to separate out the processes using an understanding of what the organization is all about, an understanding in particular of what business it is in. In the early chapters we shall look at how to model an individual process without in the first instance worrying too much about how we know that *that* pile of activity constitutes a process, dissected along natural cleavage lines, rather than a chunk that has been as-good-as-randomly hacked out of the whole. In Chapter 6 we shall discover how to do that dissection, how to divide all the activity going on in the organization into a set of processes that has been rigorously derived from an understanding of what business the organization is in. And we shall do it in a way that yields the same answer whoever does the analysis.

> **KEY POINTS**
>
> **A process is a coherent set of activities carried out by a collaborating group to achieve a goal.**
> **The chunking of organizational activity into 'processes' must be driven by an understanding of the business the organization is in.**

Before we go further, I must say how we shall be using the words 'business' and 'organization'. We shall use the neutral word 'organization' to mean any group we are interested in: a team, a department, a company, a group of companies, a company and its customers, a nation, whatever. I shall (try to) restrict my use of the word 'business' to mean 'what the organization gets up to':

- This organization is in the business of making and marketing furniture.
- This organization is in the business of providing shared services to local hospitals.
- This organization is in the business of managing building programmes for the city.
- This organization is in the business of collecting fines imposed by the courts.

Deciding on the 'organization' we are concerned with can itself be an important decision. Gone are the days when we would only worry about the efficiency of individual activities. Today we are concerned with the efficiency of, say, our entire supply chain – from one end to the other. Gone are the days when we would worry only about our patch, ignoring what happens on the other side of the wall in the world of our suppliers or our customers or our partners. Today, we need to ensure the way we work with them is fully thought through and integrated. Gone are the days when we could get on with our business in private. Today we are increasingly required to make our end-to-end processes visible.

WHY WORRY ABOUT PROCESSES?

There have been a number of business 'movements' over the last twenty years that have made people recognize that they have processes and that these processes are what the organization is about. The central notion in each is that of *process* and there is a need to be able to picture a process through a *process model*. Like all models, a process model will capture just those things that we need for our purpose. To understand the needs of the process modeller we must look at the various situations we might find ourselves in where process is important. We can readily identify the following seven. They are not completely separate but it is useful to consider them separately for now. They are:

- Situations where there is a need for a *shared understanding* of what the organization does and how it does it.

- Situations where a *common approach* to doing something is to be adopted and perhaps mandated, for instance through a Quality Management System (QMS).

- *Incremental improvement* programmes, such as might be run under the banner of TQM.

- *Radical change* programmes, such as might be carried out using the principles and techniques of BPR.

- Situations where traditional data based information technology (IT) systems need to be aligned with the needs of the organization.

- Situations in which *workflow management systems* are to be used on a computer system to control the flow of work.

- Situations where new forms of process technology such as BPMSs, are to be applied to give active support to the management and enactment of business process.

Let's examine each in a little more detail.

Understanding your organization

It has to be said that not every organization recognizes that it operates processes, even though it knows perfectly well how it is structured into functional groups and what each of those functional groups is responsible for. Whilst people might appreciate in some abstract way that the organization can only work through their collaboration, they might individually have very little idea of how the collaboration actually works – they each do their bit, but how do the bits fit together? When I have modelled a process within an organization, people will often remark 'You know, I've never really thought of things in terms of a process that starts there and ends there.' People know about what they do, who they depend on, and who they pass things on to. But they might not be aware of the larger, end-to-end process in which they, along with many others, play a part.

Simply modelling the process can provide individuals and groups with a perspective on the organization that transcends parochial views and, as a result, can promote a more collaborative spirit. 'Now I know why you want that, I can make sure you get it reliably.' We are interested in helping people to 'get out of the functional silos'.

A model that makes the process *visible* to the parties concerned can in itself bring great value.

Standardizing processes in Quality Management Systems

The emergence and development of the ISO 9000 series of standards led to an increased concern with how an organization goes about its business in a way that ensures quality in the products or services that it delivers to its customers. ISO 9001 (ISO 9001:2000 *Quality Management Systems – Requirements*) sets a standard for a QMS. Central to the standard is the notion that key processes should be defined in some way so that they are repeatable, measurable, and improvable. The details are unimportant here, but the message is: if you are concerned with ensuring the quality of your product or service, you must concern yourself with the processes that deliver that product or service.

Typically, therefore, an organization will describe how its processes are carried out in a way that:

- communicates the processes to those who must carry them out ('How should I do this piece of work?');
- provides the opportunity for independent assessment of the organization's conformance to the process it has laid down for itself ('Are these people doing what they said they would do?');
- acts as a basis for future improvement of the process ('We do this now but how could we do it better?').

Such descriptions tend to be lodged in some form of Quality Manual and they can take many forms, typically a mixture of text and diagrams. A good description – model – of a process will be one that *communicates* in sufficient detail to those that must carry it out, that is *precise* enough to permit an assessment of conformance, and that is appropriately *detailed* to be a basis for analysis and improvement. (Note the careful use of the words 'sufficient', 'enough' and 'appropriately'.)

Incremental improvement

It has long been recognized in the disciplines of quality management and TQM that the cost-effectiveness and profitability of a process are determined by the quality of the goods or services it produces, and that that quality is itself determined by the process as well as the inputs and the workers. In particular, if we want to reduce wastage (of materials, resources, or time) we need to address the 'common causes' (to use the jargon) of defects, and this means removing systemic errors, i.e. those introduced by the process itself.

To do this, we need a way of exploring at an appropriately detailed level just what happens between the customer making a request and the customer going away satisfied with the goods or services we have provided, and, within that flow, we need to understand where defects and/or unnecessary delays are introduced so that we can adjust the process to remove their cause. Central to that exploration is a model of the process, through which we can ask where improvements can be made.

The aim is that, bit by bit, we refine the process and gradually eradicate those systemic causes of poor quality: we are in the arena of incremental improvement.

Radical change

In the world of BPR, incremental improvement is not enough. Here we are looking for major breakthroughs: an 80 per cent reduction in cycle time not 10 per cent, reducing staff levels to one-fifth, not by one-fifth. And to do this we are prepared to make radical changes, not just tinkering with the fine detail of our processes but making major changes in our organizational structure, and questioning the very need for doing things the way we have done them for years, or even why we have those processes at all. We are prepared to ask questions like 'Can we operate without a central Purchasing Department?', 'Is tendering the only way we can ensure the best price for bought-in goods and services?' or 'What would happen if suppliers were paid by the recipient of the goods rather than Accounts?'

In this context, detailed maps of our current processes are largely irrelevant. We've decided that we shall only consider big changes – detail is simply not interesting. But an architectural view of our processes, and broad-brush models of the way our organization operates, of what processes we have and how they traverse the functional silos, could give us clues about the sorts of radical change we might imagine. We could question whether we could remove entire processes by thinking about how we do things in a more radical way.

And when we have decided how we want our new organization to land when we have thrown the existing one in the air, we shall need some way of designing the new processes, ensuring that they fit with those that survive and with each other, and that they make sense in our new flattened or process-oriented organizational structure.

Process design means, again, being able to model the process, this time the *new* process.

Building on database management systems

I once worked with an IT department that had built a system designed to support the information needs of a particular group within the organization. The system was designed to provide that group with a way of recording and tracking progress on the items they were processing in real time. But, at the time that the system was designed, no one had recognized that the work of that group was inextricably linked with the work of another group who had their own IT system; the new system quite simply 'clashed' with the process by which each work item had to be handled. As a result, the users resorted to inputting information *once the process had finished*, and as a result the system failed to provide the real-time support which it had been intended to provide.

Rightly or wrongly, past IT systems have often been considered to have been failures in that they have not brought the benefits that were promised to the organization. There has been an assumption in the minds of those that build IT systems (and I have been amongst them) that data and information are central: that if we start with an analysis of the information needs of the individual in the business process, we shall build a system that supports the organization effectively. Unfortunately, this ignores one important feature of organizations: they do not work simply by ensuring individuals have information at their fingertips; they work by having processes in which groups collaborate effectively. Good IT follows firstly from an understanding of the way that the organization does its business with the structures it has, and only then from an understanding of the information that the organization needs because of the way it chooses to do its business.

Process precedes information. Once we have decided how we shall carry out our business – our process – we can identify the information needs and hence the information-based systems needed to support them.

Building on workflow management systems

In the 1990s, a new class of software infrastructure products emerged: workflow management systems. These provide active support to a simple business process by controlling the flow of a work item around the organization, routing it and its supporting information and (electronic) documents and images from person to person in the process, from workstation to workstation.

Such systems clearly need some model of the process, a model that describes the path the work item takes from role to role, the decisions, the alternative paths, exception handling, escalation paths, and so on. Once again, we see the need to be able to model a process in terms of roles, activities, decisions and flows from role to role (what we shall generally refer to as *interactions*).

A key feature of such infrastructure products was that the process was pretty much set in stone: once programmed into the workflow system, changing it became a major undertaking, on a par with changing the structure of a database supporting a process: not something to be undertaken lightly and something only to be done by the experts in the IT department. Alas, workflow management systems have also tended to use proprietary models (i.e. they are all based on different process constructs) and are not general enough to express all the sorts of process we might wish to run.

Building on business process management systems

More recently, we are seeing an entirely radical class of products: BPMSs.

A BPMS takes a description (model) of a set of processes, and *enacts* it: we might say that it 'executes' it, or 'carries it out', or 'runs it'. In the same way that a normal computer 'carries out' a software program, so a BPMS 'carries out' a set of processes. This much they have in common with workflow systems. But in a true BPMS a radical step is taken: in the parlance, processes become 'first-class objects'; so within a BPMS, a process can be itself managed, revised and passed around. In this new paradigm, processes come out of the shadows and become true and visible business assets.

We can express this succinctly by saying that a BPMS supports us working *in* a business process, and supports us working *with* a business process. In the first case we enact the process, in the second we manage the process. The moment we recognize that the management of a process is itself a process, we realize that the BPMS is a new sort of world where the unit of currency is the process not its data, a world in which that currency is both minted and used – processes are defined and enacted. To use more of the parlance, we say that processes have *mobility*: rather than being something static – statically defined and statically followed as in traditional workflow systems – we allow them to be the subject matter of other processes, we allow processes to be passed around for enactment, and we allow a process to grow as a network of interacting and collaborating parts.

The modelling situation here is an order of magnitude more complex – the problem is the same as that addressed by the IPSE 2.5 project that I mentioned earlier and that contains the seeds of Riva. As a result, Riva has the necessary concepts and machinery for those modelling processes using languages such as Business Process Modeling Language (BPML) and Business Process Execution Language (BPEL).

THE RIVA METHOD

Whatever our reason for taking an interest in our processes, we shall need a way to define, record, discuss and analyse them, and we shall need a language for talking about processes. This is where Riva comes in.

Riva is a method for the elicitation, modelling, analysis and design of organizational processes.

It uses two languages for talking about processes:

- The Process Architecture Diagram (PAD) is used to describe the overall chunking of the organizational activity into individual processes.
- The Role Activity Diagram (RAD) is used to describe an individual process. In a modelling project, we shall always have one PAD and one or more RADs.

The Riva method includes techniques for:

- determining what processes an organization must have in order to be in the business it is in (chunking);
- 'discovering' and modelling an existing process;
- defining an existing process;
- designing an intended process;
- qualitatively analysing a process once a model has been produced;
- using process models for requirements definitions for information systems and workflow systems;
- developing process models for BPMS development.

In any process modelling *method* we shall want to find more than just a notation, more than a way of drawing pictures: we shall want to find 'intellectual machinery' that helps us to think about our processes and get to answers. Riva is a rigorous and rather uncompromising method: my views are that: a sloppy model cannot be relied on; a sloppy model can give the wrong answers; a sloppy model does not give real insight; a sloppy model cannot be used as the basis for any sort of computer system that is intended to support the organization, whether it be data-, workflow- or process-based.

But before we look at the notations and the techniques themselves, we must pause and examine our motivations and needs in a little more detail.

Chunking

When we set to work on an organization's processes, our first problem is getting our arms around the whole thing. Walking into the building where the business is done, we find a mass of activity going on. How on earth do we start? And where do we start? Presumably all this activity falls into some sort of chunks, chunks that make some sort of business sense. But how do we chunk it in a way that makes business sense?

Let's go a step further. If our organization changes its structure, surely we wouldn't expect to end up with a different set of chunks? It is still in the same business after all. How it does those processes might change, but it still must have those processes. And surely, to be in a particular business we need a particular set of processes? And if our organization changes its culture, surely it will still have the same processes? How the processes get done will – of course – change if we change the culture. But the existence of the process will not – it remains there, in one shape or another, as long as we are in the same business.

We shall be looking for a method of chunking that gives us a 'process architecture' with this property of invariance. What we are looking for is a chunking that is derived solely from an understanding of what business the organization is in. We would like to say 'If the organization is in this business then it must have these processes.'

When pharmaceutical drug compounds are being developed, small batches are made for clinical trials using general-purpose pilot plants. Such plants are expensive to build and run. There is a queue of batches waiting their turn for use of a pilot plant. If that queue is not managed properly, the effect on the drug pipeline can be significant: it doesn't matter how quickly each batch goes through if the important batches are held up because of bad planning. Turn to the Procedures Manual of any organization and I can almost – almost – guarantee that it will cover only 'coal-face' processes, processes that deal with *a* manufacturing batch, *a* purchase order, *a* sale, *a* customer, or *a* clinical trial. But what about that flow management? When individual cases compete for resources, management activity is necessary to decide priorities, allocate and schedule resources. We can draw two lessons. Firstly, when we chunk all the organizational activity, we must not forget that that management activity is out there and deserves equal recognition with the coal-face processes. Secondly, we cannot mix the description of how that management activity is done with the description of how an individual case is done – they are different processes.

So of all that activity in the building, we must be sure to cover both 'coal-face' processes and management processes. But what about all those people in the conference room, standing around the whiteboard discussing how they should respond to recent developments in their competitors' offerings; the finance folk sitting with their bankers looking at their exposure to currency fluctuations and deciding whether they have the right mechanisms in place to deal with them; the Board discussing possible take-over targets with their management consultants? None of these people are carrying out the day-to-day business processes, or the processes that manage those processes. They are standing back and taking a longer view of the business. They are involved in what we might categorize as 'strategy work': the results of their deliberations will change what happens and perhaps how it happens. When we put our arms round the organizational activity and chunk it up, that strategic work must also appear if we are to have a complete picture: we shall expect to find strategy processes.

In summary, we can expect our chunking to expose processes in three flavours: coal-face, management, and strategic. We shall examine these more thoroughly in Chapter 5, before describing in Chapter 6 how to use them when developing a process architecture.

Describing, designing, analysing, enacting

A recurrent theme in this book is the idea that what is included in a model for an individual process and what is not will be determined largely by the perspectives we choose to take, which in turn are determined by the reasons we have for modelling in the first place. We have already seen that there are a number of reasons for taking a process-oriented view of our organization, and that we have many reasons for modelling the processes within it; four Ds and an E: discovery, definition, diagnosis, design, and enactment.

Modelling to discover and define a process

This is what we do when we want, amongst other things, to:

- discover a process that has not seen the light of day: 'This is how we apparently handle customer complaints';
- define a process: 'This is how we will all handle customer complaints';
- communicate it to others: 'This is how your work contributes to customer complaint handling';
- share it across a group of people: 'So this is how, together, we handle customer complaints round here';
- negotiate around it: 'If you could do this, my life would be made much easier; in return I can ... '.

It is surprisingly common for an organization not to have a clear idea – or sometimes any idea – of how certain things are done. They do get done because people fit in, work things out, develop their own patterns of behaviour and pass them on. There is a sense in which this could be regarded as 'good': it promotes flexibility, discourages slavish subservience to a set of rules, allows things to develop and shift as the environment changes, and so on. On the other hand it becomes hard to promote good practice, makes life hard for new recruits, doesn't permit steady improvement, allows reversion to bad practice without anyone noticing, and so on. Process modelling often has a role to play in revealing to the organization how things are, perhaps how bad things are.

We might call this 'process discovery'. It is about building a shared understanding of the process as it is today, an understanding that we can communicate to others.

Such discovery might be a first step to agreeing on a common way of doing things. Suppose a group is seeking ISO 9001 certification for its QMS; it will want to define its processes in its Quality Manual. We shall probably find a descriptive process model in some form in such a Manual. The model acts as a work instruction to people in the organization. Text is very often used to describe how things are done but the serial nature of text makes it impossible to adequately describe – let alone prescribe – something that has possibly many threads, decisions, concurrent activities, and so on. A diagram is a traditional way of dealing with this.

A software house had a number of its processes described in its Quality Manual in the form of RADs. Those diagrams told people what was expected of them when carrying out the processes; in particular they specified in quite considerable detail the key business processes of planning and reporting projects, of bidding for new contracts and of purchasing – all processes that had an important and direct financial impact on the company and to which it therefore wanted to ensure some degree of conformance. Other processes were about delivering product quality. RADs allowed them to define all these processes to an appropriate level of detail.

Summarizing our needs for process description is difficult: we can have so many different reasons to describe processes. Certainly our model must say what we want it to say. So the notation must allow us to say those sorts of things, and our method must help us do that. We shall look more closely at the practicalities of modelling for discovery and definition in Chapter 9.

Modelling to design a process

It is relatively rare for people to have cause to design a new process or set of processes from scratch. It probably means that either a new organization has been created or an existing one has undergone a very radical change of business.

One group of managers in a multinational product company wanted to define a planning process that they would use involving their management and other corporate functions such as 'IIQ', 'Finance' and 'Audit'. They already had a process in place but it was not well articulated and they were fairly certain that they did not all have the same view of the process. The process was that of deciding on the portfolio of products in which they would fund investment, and the process of developing a new product within that overall planning process. They already had a short verbal description of the proposed new process and we used the methods described in this book to capture those processes completely. In addition to giving them greater understanding through the analysis, the method gave them a clear diagram which then went into their Procedures Manual. It also filled in many gaps in the textual description, which only became apparent when the process was explored using Riva's rigorous notations and method.

With the fragmentation of industries such as utilities, and the consequent construction of new regulatory frameworks and bodies, entirely new processes become necessary to ensure fair competition between the fragments. These can be truly greenfield sites for process design. Nothing similar exists in the current organization; entirely new objectives arise, and new roles and responsibilities must be created, and so on. A new utility organization found itself having to set up a new group with responsibility for reporting to the regulatory body. Nothing like this had been done before: what processes would they need? Never having had the need before, there was no organizational structure to support it. How should they design those processes in the absence of the 'comfort' of an existing organization?

We shall look more closely at the practicalities of modelling for process design in Chapter 11.

Modelling to diagnose a process for improvement

Once we have a model of an existing process, we may well want to use the model to analyse the process itself. Some of the questions we shall want to ask will be quantitative: 'What is the average cycle time?', 'How much will the cycle time be affected by changing the process in this way?', 'Where are the bottlenecks?' Other questions will be qualitative: 'Do we have the optimum division of tasks across the people involved?', 'Why does this paperwork flow back and forth?', 'Are the right decisions being made at the right level in the organization?', 'Are we overdoing the financial oversight?'

Such analysis is a common precursor to improving the organization by, for instance:

- changing the order of activities;
- changing responsibilities for activities or decisions;
- changing the way things are scheduled;
- increasing or decreasing the amount of parallel activity;
- removing or inserting buffers or stores for materials between steps in a process;
- restructuring functional groups to align them better with the process.

Any organization involved in TQM or other process improvement activity will want to model its processes and analyse them for weaknesses or inefficiencies on paper, before trying out improvements in real life.

We shall look more closely at the practicalities of modelling for process improvement in Chapter 10.

Modelling for traditional information systems

Traditionally, when considering computer support for the activity of an organization, we have done some sort of analysis of the organization in order to identify where automation can best be applied. Since the mid-1970s that analysis has concentrated on data and into the twenty-first century data analysis continues to be the cornerstone of software development: object analysis, data-flow analysis, Entity Relationship (ER) modelling, and Entity Life History (ELH) modelling in particular have featured large and continue to do so. The move towards object-oriented conceptualization has changed that scene little.

Unfortunately, too little attention has traditionally been paid to understanding the business *process* that is being supported. Even an otherwise well-thought-through system development method such as Structured Systems Analysis and Design Methodology (SSADM) – see for instance *Business System Development with SSADM* (Office of Government Commerce, 2001) – has taken the view that business activity is relatively unstructured, or at least that any structuring is not of great interest. With the advent of the Unified Modeling Language (UML), we have seen an increased interest in the nature of organizational activity, with the idea of so-called *use cases*.

We can look to our process modelling method for two things in the Information System (IS) context. Firstly, we can expect it to reveal more effectively, reliably and efficiently the use cases of the business. Secondly, we can expect it to make the information needs of the individual clearer in the context of the entire process. We shall look more closely at the practicalities of modelling as a precursor to defining requirements for traditional information systems in Chapter 12.

Modelling for workflow management systems

Traditional information systems have for the most part had a very simple architecture. In essence, they provide the individual with a peek 'n' poke facility into some central database – most importantly *they only give explicit support to the individual*. Every individual is plumbed into the data. But organizational processes are not just collections of individuals operating independently. Organizational processes are carried out by groups of people acting collaboratively to achieve a goal. If our computer systems are to reflect this they must support groups more directly. In the last two decades software packages have appeared under the title Workflow Management (WFM). Their architecture recognizes that people who connect to the system are working jointly on a case, and so that architecture must support the flow of a case from person to person. They have recognized that organizational activity is more than a set of individuals manipulating data, and is better seen as a flow of work items between collaborating individuals.

Now, traditional data-oriented systems analysis and design have served us perfectly well when we build data-oriented systems supporting individuals. But WFM products require different analysis and design methods. Data-oriented analysis methods provided us with various sorts of data model. WFM products need process-oriented methods providing us with process models.

We shall be looking for ways of identifying workflows and then understanding the nature and content of that flow; in particular, who gets to do what when, what information they need to do it, what information flows with the workflow, and what is static. We shall cover this in Chapter 12.

Modelling for business process management systems

Given a data model for a data processing system, any Database Management System (DBMS) will allow us to store this in a database and use it directly to generate forms and reports that the individual can use to add, amend or present data. There is a thin sense in which we could say 'The data model is executed.'

What's the analogue for process models? Suppose we could give our process model to a computer system and have it *enact* that model, i.e. 'run' the model, supporting the participants in the process as the process proceeds, handling their agendas, supporting their interactions, and perhaps playing its own part in the process. Systems that provide this sort of support are termed *enactment systems* and they provide us with our final motive for process modelling: they require a process model whose 'meaning' is sufficiently well defined to allow them to enact the process without further human intervention.

Such an enactment system will go beyond handling simple workflows and will deal with the network of processes that can be in progress at any one moment. Processes themselves will become a sort of currency in the system: they can be changed and passed around. This use of process models will have important implications for the process modelling method and its notations. But we can imagine the step beyond that we touched on at the start of this chapter: the idea that our processes, 'living' inside the enactment system, can themselves be manipulated in a variety of ways: small changes, big changes, new processes even. As well as supporting our use of our processes, such a system would support our *management* of our processes. It becomes a BPMS. Not surprisingly this introduces a whole new raft of requirements on our modelling notations and method.

We shall look more closely at the practicalities of modelling for BPMSs in Chapter 13.

KEY POINTS

Our modelling method must enable us to:
- **take an architectural view of our processes;**
- **expose our processes as a discovery activity;**
- **define our processes appropriately in a regulated environment;**
- **use our models to diagnose our processes;**
- **design new processes from scratch;**
- **expose the information needs of the participants in a business process as a prelude to building an information system;**
- **design workflows that can be supported by a workflow management system;**
- **produce the process definitions that can be enacted in a BPMS.**

EIGHT PRINCIPLES FOR PROCESS MODELLING

To finish this chapter, I'd like to introduce a Tutor and a Pupil in discussion about some of the requirements of a process method. Their dialogue – the first of a number throughout the book – highlights some important principles that underlie Riva.

Principle 1: If we must have abstractions, let's make them meaningful

Tutor: When we model something we describe it, and to describe something we need a language. A model uses a limited language – a limited number of concepts – that allows us to say the things we want to say, to describe the things we want to describe. If we wanted to model a nation's economy, how would we model it?

Pupil: We might capture it in terms of the flow of money between the various places it can reside: the Treasury, people's savings, money in circulation, investment instruments of various sorts, and so on?

Tutor: Yes. Note how we'd be working with two abstract concepts: a 'pool', of which each of those residences is an example, and a 'flow' – the movement of money between two pools. By varying the rates of flow we can investigate the behaviour of the model and perhaps deduce something about the behaviour of the real economy. An early economic model used precisely these abstractions, and made them concrete by using plastic containers of water to represent pools, and piping and pumps between containers to represent flows. The pools and flows were represented as plumbing and the money by water.

When we model processes – what people do – we shall need a small number of concepts that represent real-world things, but they must be concrete enough for someone looking at our model to readily understand the model and what it is telling them.

Principle 2: The real world is messy

Tutor: At a suitably high level of abstraction, any process can be made to look neat and tidy.

Figure 0.1 summarizes the process of collecting taxes from you and me and giving them to the government to spend on our behalf. Tidy isn't it?

Pupil: Well, it certainly summarizes what tax collection is about: namely that there is some relationship between The People and The Government that involves The Tax Collector.

Tutor: So how is it that tax collection causes so much grief and there is so much money to be made out of helping people with their tax affairs? The fact is of course that the tax collection process is complex and involved. A summary diagram hides all that – and can make it next to useless if we want to answer any questions.

FIGURE 0.1 *Tax collection: pure simplicity*

Pupil: Could we 'decompose' this neat picture in some way? We could still keep it neat by ensuring that we restrict the number of boxes on the diagram to, say, no more than seven. After all, someone did once say that people have trouble with handling more than seven items of information at any one time. And, if we want to, we could allow ourselves to 'decompose' any one of those boxes in turn into its own diagram with another seven-ish boxes to describe it, and so on.

Tutor: You seem to be saying that our process model can take the form of a hierarchy, consisting of a set of diagrams, each of which (with the exception of the topmost one) expands a single box on its 'parent' diagram. Something like Figure 0.2 perhaps? A neat hierarchical model? OK, then answer me this: if this is a 'good' model, then it must be 'like' the real world it claims to model. Is it? In particular, does the structure of the process in the model reflect the structure of the process in the real world? When we look at the world, do we find neatly ordered processes, with each activity in a process neatly decomposable?

Pupil: I guess not. I suppose if we are looking at something statically – a city for instance – we could draw a model of its static nature and say that the city is made up of quarters, and each quarter has districts, and each district has streets and so on ... But we wouldn't be saying anything about the way that the city works.

Tutor: Exactly, and your point about the difference between a static and a dynamic view is very important. When we model organizational activity we are interested in dynamics. Dynamics are not hierarchical. Organic systems are networks, often networks of networks. And they are changing networks: things are born, establish relationships with other things, and then die. The network is a dynamic thing, constantly changing. There is a dynamic flux of activity ... not some static hierarchy of activity.

Pupil: Is there another sense in which models can be complex: things can simply get ... messy? In fact I guess we should be very surprised if something like the process of an organization that has grown organically over time was so neat and tidy.

Tutor: True. We'll see that – when they happen – real-world processes can be complex, gangling, even muddled or messy (these aren't my words – they're words frequently used by the people whose processes I work with). They wander here, there and everywhere; the hierarchical picture fails to show that: it replaces convolution by neat (but irrelevant) hierarchy. The message

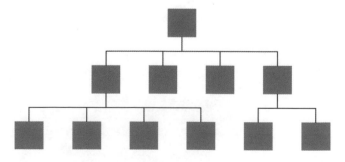

FIGURE 0.2 *Are the world's processes really this neat and tidy?*

for our process modelling approach is that we can't let the notation dictate to us that our models must be neat hierarchies. Things are much more complex.

Unfortunately the IT world has for years been using hierarchies as a way of designing synthetic things like computer programs, and software structures of all kinds. The great mistake has been to assume that we can use such hierarchy-based notions to describe *organic* things like businesses. Businesses aren't the product of tidy design activities over which some designer has had full control. They're the product of time and many uncoordinated changes by many hands. Traditional structured approaches from the IT world just won't work.

Principle 3: A model must mean something, but only one something

Tutor:	What does Figure 0.1 mean to you? What do the two boxes mean? What does the arrow mean?
Pupil:	Errr … from their names, the boxes seem to be collections of people … or one is a collection of people and the other is an organization …
Tutor:	I haven't told you what a box means, of course. 'Money' doesn't appear anywhere in the picture – are you surprised?
Pupil:	Well, I assumed the arrow represented money.
Tutor:	It could. But it could also represent 'paperwork'. Suppose we accept that the arrow represents money flowing from The People to The Government. Can we assume from the single head on the arrow that the money only flows one way? In fact it doesn't. I recently got a tax refund. Would having a two-headed arrow be more accurate? Let us assume that, because the arrow has the label 'The Tax Collector', the Tax Collector is the agent for the flow, the thing that makes it happen on behalf of The Government. Would that mean then that the Tax Collector sometimes makes the flow go from The Government to The People? Does the Tax Collector act as your agent for giving money to you?
Pupil:	It was such a simple diagram – two boxes and an arrow – but it does seem to raise more questions than answers.
Tutor:	Right. And the reason for that is that I have not given a clear meaning to the symbols I've used in the diagram. 'A picture says a thousand words' – a different thousand words to each of us, unless we agree on the semantics of our notation. If I can't tell what my process model means, I can't tell what needs doing. When I choose a notation for my models, I need to be sure that I have clear semantics for them: I know what a box means, I know what a line means, I know what each little symbol means. The model is unambiguous.

Principle 4: Process models are about people, and for people

Pupil:	I can see a danger here. When we model a process we are describing what people do. (I suppose machines of various sorts might play a part in a process, but they don't have to be asked their opinion so I guess we can concentrate on people.) To find out what people do we'll need to ask them

and watch them, we'll do some sort of 'process elicitation'. I'm sure we'll then want to put our model of their process in front of them and have them discuss it, correct it and improve it and ask whatever questions of it they want answered.

So the concepts we use in our model must be concepts people relate to in daily life. The notation has to make sense to people. If we can't explain it in ten minutes, it doesn't make sense. But on the other side of the coin, any well-founded notation will have subtleties and potential complexities, and so there's a danger of opacity. A newcomer to a model must be able to understand quickly what the model is saying. If the model needs extensive interpretation by a skilled analyst before it can be understood by the person on the shop floor, the battle is half lost, surely.

Tutor: Yes, there is a dilemma there. We want a notation that is powerful – that allows us to say all the complex things we want to say – yet is accessible. The acid test is this: at their first modelling session, can an ordinary person go to the whiteboard and correct a mistake in the model so that it correctly describes what happens?

Principle 5: There's what people actually do, and there's what they effectively do

Pupil: I have another problem. When the Accounts Department raises an invoice against a customer's order, what we see them doing is involving themselves and others in a paperchase, in which the flow of paper is embellished with activities of transcription, checking, updating, copying, chasing, phoning, and so on. But what they're *effectively* doing is extracting payment for the goods and services supplied. How do we reconcile these two views of a process?

Tutor: The answer is that we won't try to reconcile them: we'll recognize that these are two different views of the same thing. When we look at a process in concrete terms, we see what people actually do, and such a model might be in terms of the *mechanisms* they use. But, when we want to understand how the process might be re-engineered, or what the possibilities for technology support are, or how the organization and roles and responsibilities might be changed to improve things, we shall need to get to the bottom of the process, to understand what it is *effectively* about. We would draw a different model that expressed the *intent* of their actions.

Our modelling method and notation must allow us to prepare both sorts of model, what we might call *concrete* and *abstract* models.

Principle 6: People work in functions, but they do processes

Tutor:	If you ask someone what they do in an organization, what sort of answer will you get?
Pupil:	'I work in Accounts' or 'I work in Product Design' or something like that.
Tutor:	And if you press for detail?
Pupil:	They'll tell you how they contribute to the work of the department, where the work comes, from and what happens as a result of their work.
Tutor:	Right. The lessons from TQM and BPR have been that what people really do is play a part in one or more *processes*, and that these processes invariably cut through the boundaries between departments. Getting the organization to recognize the existence of these 'cross-functional' processes and to deal with the conflict that occurs at the boundaries between the 'functional silos' is a major part of improvement and re-engineering.
	There's a big message here for our modelling approach: do I have ways of modelling the process from the point of view of organizational structures and how work is allocated across the functions, as well as from a pure responsibility point of view? Can we separate out responsibilities from organizational functions? We have to be able to do this if we are considering re-engineering through change in the functional structure or in the responsibilities allocated to functions. Remember that we can have a functional group called, say, Administration, and then have to decide what responsibilities we want to give them.

Principle 7: It's what people do, not what they do it to, that counts

Tutor:	What do you do?
Pupil:	I ... carry out activities: I write an article; I facilitate a workshop for a client; I elicit a client's process; I give a training course.
Tutor:	Just one thing at a time?
Pupil:	Well ... I often have a number of activities going on in parallel: I start a new piece of work for a client; I put one activity down to do another; I stop everything to fill in my timesheet; I resume one of the activities I put down; I finish an activity.
Tutor:	Right, so you have a number of balls in the air, a number of plates spinning, a number of activities you could be doing, or are in the middle of doing, at any one moment. Anything else?
Pupil:	I make decisions: I decide that the text for the brochure need not go for another review but can go straight to the graphic artist; I decide that a request for new computer hardware is acceptable; I decide that an invoice is correct and can be sent to the client.
Tutor:	Yes, you're choosing among alternative courses of action depending on the current circumstances. More?
Pupil:	I interact with other people: the graphic artist and I agree on the final layout of the brochure; I send the manuscript of my book to the publisher; I delegate responsibility for hardware procurement to my facilities manager.
Tutor:	OK. Let's pull this together. In all these things that occupy your day, you're doing things, sometimes on your own, sometimes with others. You're

playing a part in a number of different processes at any one time, and in each you may have a number of actions in progress. It's reasonable then to expect that any notation we use to model those processes should allow us to capture activity: concurrent actions, collaborative interactions, and decision making.

We must focus on what people do, collaboratively and individually. What they do it to is of secondary importance.

I raise this because here again there has been an unfortunate influence from the IT world on our process modelling (and I speak as a software engineer of many years). Hitherto the IT world has concentrated almost exclusively on *data*. On what people do things to. This is not surprising. Computers gave up computing pretty soon after their invention to concentrate on looking after people's data: memory became a cheap, voluminous commodity and that memory was as good as permanent. We've exploited that. When IT people build information systems they principally design databases and ways of getting the data in and out. The traditional development methods (SSADM, IE, Yourdon, and so on) rightly concentrate on the data aspects of the business being supported, and the notations are about data: Data Flow Diagrams, Entity Models, Entity Life Histories etc. But we should not expect their hammers to be good for driving home our screws. We are in the process business not the data business.

KEY POINTS

- If we must have abstractions, let's make them meaningful. Any process modelling notation must deal in business-oriented concepts that people relate to. Otherwise how can they tell if a model's right?
- The real world is messy. The notation must be able to model mess when necessary. Muddle modelling is perhaps the norm, not the exception.
- A model must mean something and only one thing. If our model is ambiguous, how can we tell what it is telling us or others?
- Process models are about people, and for people. The notation must make sense to people. If we can't explain a model in ten minutes, it doesn't make sense.
- There's what people actually do and there's what they effectively do. These are different and we must be able to model both.
- People do processes, but they work in functions. These two can be in conflict. A model must capture both – and the conflict.
- It's what people do, not what they do it to, that counts. A process is principally about doing, deciding and cooperating, not data or things.

THE STRUCTURE OF THIS BOOK

The book falls naturally into two parts: you can think of part 1 as theory and part 2 as practice.

When we want to think about a particular subject in life we need an appropriate vocabulary – words to describe the subject – and a grammar – ways of arranging words to convey meaning. If we choose the right words and the right grammar we shall be very expressive. Major branches of mathematics were held up until a notation was developed that not only represented the subject being worked on, but also worked as a notation. The same will be true for our subject: processes. If we choose the wrong vocabulary or the wrong syntax, we shall not be able to say things we want to say, and we might even end up saying things that are just plain wrong; we shall be describing processes that don't exist; or we won't be able to answer the questions we want answered.

So part 1 is very much about getting the right vocabulary and the right syntax so we can describe business processes in a way that meets our needs, whatever they may be.

- Chapter 1 – Basic process concepts – gets us thinking about just *how* we look at a process. What are the features of real-world processes that we want to reflect in process models?

- Chapter 2 – Modelling a process – provides all the 'vocabulary' necessary to represent a single process in a RAD. We shall examine the notation and the underlying concepts in detail.

- Chapter 3 – Dynamism in the process – highlights the levels of within-process concurrency that can be captured in a RAD, and shows how we can exploit that richness when we model real-world situations, current or planned.

- Chapter 4 – Process relationships – examines the types of dynamic relationship that processes can have and illustrates how we represent them on RADs. The relationship types will be central to the construction of our process architecture.

- Chapter 5 – The three basic process types – describes the three main types of process – the Case Process (CP), Case Management Process (CMP) and Case Strategy Process (CSP) – that underlie the construction of a process architecture.

- Chapter 6 – Preparing a process architecture – deals with the construction of the *process architecture* of an organization, a concept of central importance for re-engineering, for overall process design and for steering any process work.

- Chapter 7 – Dynamism in the world – shows how the process architecture captures all the between-process concurrency in the world.

Part 2 puts part 1's theory into practice.

- Chapter 8 – Managing the modelling – provides guidance on running a process workshop and conducting interviews in order to prepare a model of a process, for whatever reason. We shall concentrate on how to make appropriate modelling decisions, the need for fitness for purpose in process models, and how to get results quickly. Subsequent chapters customize this general approach for different purposes.

- Chapter 9 – Discovering and defining processes – covers the practical use of the approach in determining what processes an organization has, in eliciting those processes onto RADs, and in the use of RADs in QMSs, tying into ISO 9001 with its emphasis on process.

- Chapter 10 – Analysing for process improvement – is about using the approach at both the architectural and the process level for asking questions about processes and their performance, and for driving tactical process improvement.

- Chapter 11 – Designing a process – covers the design of a new process architecture and new processes. As the processes do not exist today, we shall start from a blank sheet of paper.

- Chapter 12 – Processes and ISs – covers the use of the approach in constructing an IS strategy for an organization, and in the design of ISs.

- Chapter 13 – Processes and process systems – covers the use of the approach in using the new wave of BPMSs in which agile and mobile processes replace static data structures.

WARNINGS

Some warnings are in order.

Riva is not just diagrams. We shall be drawing diagrams – three sorts in fact – but they are only half the story. The other half is the underlying concepts and how they can be used to get a real understanding of complex human organizational activity. Without them, the diagrams end up just being sequential flowcharts and the point is missed. If you feel tempted to simply check out the different sorts of blobs on a RAD in Chapter 2 and start drawing, you will be missing a great deal.

When you come to the section on the RAD notation, you will find yourself reading (what I hope are) very precise definitions of things, with apparently simple concepts being teased apart mercilessly. You will ask yourself why all this is necessary. If you want rough and ready models, you will draw rough and ready models. You will misuse the Riva notation, ignore the subtleties and add new blobs and arrows of your own. In my defence, I must tell you that one reason for drawing a model of a process may be that *we want to execute the model*: we want to give the model to a computer and have it run the process for us, with human beings playing the roles and carrying out the activities and interactions, all under the control of the machine that has the process in front of it. Drawing a few blobs and arrows on a whiteboard can be a rewarding experience on its own: things become clear, relationships are exposed, and we can come to a shared understanding (we like to think). But those blobs and arrows do not capture the process with the *precision* required to give that 'model' to a computer to execute for us: a goal of BPMSs. Chapter 13 will take us into that more refined world, where slapdash will not be sufficient, and where precision rules. When you use Riva you have the opportunity to be very precise, whether or not you choose to take that opportunity.

Riva is a method for the analyst. This being so, we shall not be afraid of using precise and specialized terminology between us as analysts. This book introduces a number of detailed technical ideas and terms, essential to the analyst for real understanding of a process, and for the accurate capture of a process in a process model. But one person's terminology is of course another's jargon, and the analyst needs to be careful when working with 'ordinary' people. My (good) experience is that, whilst I have these technical ideas and terms in my head, ordinary people can work with Riva process models happily and productively without them. My (bitter) experience is that the effect on an ordinary person of hearing the word 'instantiate' for example is akin to a sharp blow between the eyes with a heavy club: it switches them off. Exercise caution.

Like all methods Riva does some things and not others. It is a set of ideas. There are many ideas for different situations. There is no obligation to use all of them at the same time, only to pick the ones that you need. You can use Riva in a rough and ready fashion, or you can exploit all the subtleties and precision it offers.

Riva is not a cookbook. There is no recipe and there are no cooking instructions. There are concepts to be used. You choose.

SOME EARLY REFERENCES

Holt's original exposition of Role/Activity Theory (Holt, Ramsey and Grimes, 1983).

Greenspan's thesis on his Requirements Modelling Language (Greenspan, 1985)

Clive Roberts's and my paper on the IPSE 2.5 work (Ould and Roberts, 1987).

1 **Basic process concepts**

What are the features of real-world processes that we want to reflect in process models?

WHAT HAPPENS IN THE WORLD?

Our analysis of the needs of the process modeller leads us naturally to the basic concepts that we will be looking for in our process modelling language – the vocabulary that will help us say the things we want to say, to answer the questions we want to answer. This chapter will look at those basic concepts, before we go on to model them in Chapter 2.

Processes are divided over roles

Central to our notion of a process is that it is about *people* doing business – how they do it, how they think they do it, how they are supposed to do it, how they might do it better or differently, and so on. People are central. So should we model a process by saying what Mavis does, what Bill does and what Charlie does? Perhaps we should draw a box labelled 'Charlie' and in that box describe, in some way, what Charlie does. Similarly we might have boxes for Mavis and Bill, and for the other people involved in the process. This doesn't feel right of course because, in an organization, people do things not because they are themselves but because they have a *responsibility* in the organization; they are perhaps paid to carry out that responsibility: they have a *role* in the organization. So those boxes on our diagram should really be about roles, responsibilities that are carried out – *acted*, as we shall say – by individuals or perhaps groups of individuals.

For now, let's just think of a role as a responsibility of some sort, perhaps a hat that someone – an *actor* – can wear. I go to work, and as I walk in the door of my office I put on my hat of responsibility.

In the process of getting a book published we find roles such as the author, the publisher's editor, the typesetter, the copy editor, the publicist, the printer, and the bookseller. Each has their own responsibility.

Some roles are carried out by individuals: Martyn Ould is (carrying out the responsibility of) the author of this book. Other roles might be carried out by a group of people: the responsibilities of the Accounts Department are carried out by the people who work in the Accounts Department.

Finally, a role has the things necessary to do its actions. They can include its physical environment, work in progress, materials, resources in various forms, and tools. We shall call these the role's *props*. Those props might reside permanently within the 'body' of the role; for instance, the Project Manager has a set of plans to use during the project. Or props might be passed to the role; for instance the Project Manager receives terms of reference for the project to be managed.

Individuals do things following rules

In carrying out my responsibility – in performing my role – I do things. We shall call those things *actions*. So roles carry out actions.

In the business of getting a book published, various people write the book, prepare the index, draw the diagrams, check the copy, set the type, print the book, and get copies to the shops. By predicting the target market place, the publisher decides whether the book will come out first in hardback or paperback. The designer decides on typographical issues such as layout and typeface. The printer prints the pages and binds the books, ready for the retailer to sell to the public. Each of these actions does something in a way that we hope adds value and contributes to the business of achieving the goals of the process.

Things are done in a particular order: we cannot label a product until we have the label and the product; we cannot typeset a book until the copy is ready; we must lay the foundations of the house before we build the walls; we need to get the budget approved before we spend money.

Sometimes the way things are done is determined by the outcome of decisions about the state of things: if the customer is late in paying, we charge interest to their account; if the house being insured is in a flood plain, we estimate the insurance premiums differently; if it is the weekend, we charge double; if the email system is down, we send a fax.

Sometimes, how we do things is governed strictly:

- Policy. Our company might have a policy that nobody can approve their own work, or that all product tests must be carried out by someone independent of the production group.

- Procedures. Some of our work might be regulated and defined in the form of procedures. For example, in order to control financial commitment closely we might have procedures for planning projects, reporting project status and purchasing. We might have procedures to make interfaces efficient: all requests for training follow the same procedure so that people requesting training do not need to invent how to make a request, and the people handling requests know in what form they will arrive. Procedures might say who can sign off purchases above what value, who can authorize a change in production schedules, or who can cancel a project. (Note that I have used the word *procedure* here in the strict ISO sense of standardized activity.)

- Standards. A standard might be laid down to define a common appearance or content for something produced during the process. We might require that project plans and reports conform to a particular layout. This might be for efficiency (everyone knows what a report looks like and where to find the information they are interested in) or for control (we want to ensure that certain topics always get covered and certain information is always included).

All these are *business* rules that govern 'how things get done around here'.

Individuals within a group *interact*

Also central to our notion of process is that people, in carrying out their roles, sometimes do things *together*: they collaborate. To collaborate, they carry out some actions together. We shall say that they *interact* and that they *have interactions*. For instance:

- You and I discuss something.
- You and I negotiate.
- I contract with you to do something.
- I pass you some information.
- I delegate a task to you.
- I ask you for something.
- I give you authority to do something.
- You and I agree on an action.
- You and I jointly approve something.
- You report your status to me.
- I oversee something you are doing.
- You pass me the results of your work.
- I instruct you.
- You and I work on something together.
- I wait for you to do something.
- I chase the progress of your work.

Note how rich interactions can be. An interaction isn't just about the locus of activity moving from one role to another; real business-oriented, value-adding things can happen in an interaction. Interactions are just as vital to the process as the actions that individual roles carry out on their own. Ineffective interactions can be as damaging to a process as ineffective actions. Slow interactions can affect cycle times as much as slow actions.

So, a role involves a set of actions and interactions which are governed by rules which, taken together, carry out a particular responsibility. A process is the sum of the contributions of the individuals acting as individuals and collaborating as a group. If we get the sum right, the process achieves its goal. So ...

Processes have goals

A process is done for a reason: it has a *goal*. Sometimes the goal might not be reached and there is some other *outcome*, perhaps an undesirable one, some sort of failure.

For instance, the goal of a process might be to deliver a computer system, to maintain positive cash flow, to organize the efficient use of a piece of plant, to provide a medical care service to a customer, or to manage a research budget. It must be possible to see from our process model how a process is achieving the goals set for it, and ideally to be able to identify the point(s) in the process where those goals can be said to have been achieved.

KEY POINTS

A *process* is a coherent set of actions carried out by a collaborating set of roles to achieve a goal.
A *role* is a responsibility within a process.
An *actor* carries out a role.
A role carries out *actions* following business rules.
A role has *props* which it uses to carry out its responsibility.
Roles have *interactions* in order to collaborate.
A process has *goals* and *outcomes*.

These concepts are central to Riva: role, actor, action, interaction, goal, and outcome. Let's examine each in more detail.

ROLES

Suppose we walk into a supermarket company and identify the things that we might think of as roles, in some yet-to-be-defined sense. Here are some: *Store Manager, Shelf Stacker, Checkout Assistant, Shop-floor Assistant, Security Guard, Finance Clerk, Warehouse Person.* If we walked into a publishing company we might choose *Author, Editor, Commissioning Editor, Marketing Manager, Copy Editor, Production Planner.* If we walk into a software engineering company we might find *Project Manager, Programmer, System Tester, Chief Architect, Designer, Configuration Controller, Finance Director.*

Roles of this sort are rather like job titles, the sort of thing an individual might have printed on their business card. Or they might be boxes on the organization chart. Or both. For instance, there will be a box on the organization chart labelled *CEO*. There might be several labelled *Store Manager* – one per store. Such roles have a part to play, responsibilities to carry out in the organization. However, there will be no box labelled *Author* on the organization chart for our publishing company, even though authors have a lot to do with the processes of a publishing company.

Author is not a post in the organization, but a real author might view it as their job title, what appears in their passport as their occupation, or on their business card.

Moreover, I could stand back from the organization a little and see it in terms of rather 'larger' roles: *Finance, Production, Editorial, Marketing*. Here I am clearly spotting functional *groups*, each of which has one or more responsibilities in the organization. What is the relationship between these sorts of roles and the 'smaller' roles we identified above? We might expect that the *Finance Director* carries out some of the responsibilities carried out by *Finance*, and that some other parts are carried out by (say) *Finance Clerk*. Of course, we can't divide up all the responsibilities of *Finance* and hand the bits to the 'smaller' roles because responsibility isn't like that: in many cases, a responsibility of *Finance* can only be carried out through the cooperation of a number of roles operating within the Finance Department. I am suggesting here that it is dangerous to think of roles nested in some sort of hierarchy: hierarchies smack strongly of decomposition and if we try to decompose responsibility by cutting it up into lumps we shall lose the notion of cooperation and collaboration, which is what makes so much happen in organizations. Things will fall apart.

There is probably only one Finance Department in our supermarket company. But we might identify *Store* as a role in the organization and there will be lots of them. A bank may have many branches: each branch plays the same role and has the same responsibilities.

There are other sorts of role which are neither posts nor job-titles nor functional groups. Take *Customer* for instance. When I walk into a shop, I take on the role of *Customer*, and in that role I interact with roles in the shop. I don't have *Customer* in my passport, any more than I have the role *Expense Claimant* on my business card. But there are times – month-ends typically – when I do put on the hat of *Expense Claimant* in order to claim my expenses. Both of these are *transient* roles that I take on at appropriate times. Here are some similar transient roles: *Job Applicant, Hospital Patient, Complainant, Employee, Defendant*, and *House Vendor*.

So far, our roles have been rather tangible things: you can kick them, or at least draw some sort of line round them somewhere. But we could take the notion of role-as-responsibility a step further and identify rather abstract things – pure responsibilities – as roles. For instance, *Large Claim Approval* might be a responsibility in an insurance company. When claims over a certain size are about to be paid, we might require that they are scrutinized independently before going through. The responsibility for carrying out that scrutiny has a reality – though we might not be able to kick it – and an identity. In practice, we shall probably allocate that responsibility to a role of one of our other, more concrete types, a particular post or job title, say.

Let's pull these threads together and list the different sorts of roles that we might have:

- A unique functional position or post: e.g. *Head of Analysis Department, CEO*. Such a position or post is unique in that there is one and only one in the organization. There is only one Analysis Department and that brings with it the responsibility of heading it, which we wrap up in the unique post of Head of Analysis Department.

- A generic functional position or post: e.g. *Head of Department, Divisional Manager*. Each Department brings with it the responsibility for heading that Department. Each Division brings with it the responsibility for managing that Division.

- A unique functional group: e.g. *Document Registry, Accounts, The Government*. Such a group is unique in that there is one and only one in the organization. It has a part to play in the organization's activity.

- A generic functional group: e.g. *Department, Branch, Subsidiary*. There may be many Departments or Branches. They all have the same responsibility in the organization.

- A generic type of person: e.g. *Trade Union Member, Customer, Purchaser, Expense Claimant*. Typically a rather transient role – one that comes and goes – but there may be many at any one time.

- An abstraction: e.g. *Progress Chasing*. Such an abstract role is almost a definition of the responsibility itself, rather than a label of a post that brings with it a responsibility.

KEY POINTS

Some roles are unique in the organization and some are generic and replicated.
Some roles are concrete and some are abstract.

Roles are types; role instances are acted

We have so far talked rather glibly of roles, and of roles being acted. But we need to introduce an important subtlety here.

In the world of cars there are many different types, for instance a Rover or a Saab. Given a type of car, we will find on the roads a number of cars of that type. I have a car of the Saab type; a colleague has a car of the same type; our two cars are both instances of the Saab type.

These notions of a *type* and the *instances* of the type are very important in the Riva method. In fact, a role in Riva is actually a type. That is, in principle there can be a number of different instances or *occurrences* of a role type active at any one time within an organization. In this book I shall use the word instance in this sense of occurrence. (Software engineers will recognize the terminology of object orientation.)

As an example, in a pharmaceutical R&D company – let us call it Hill Pharm – we might have a number of projects running at any one time, one for each new drug that is undergoing development. Suppose we define the role *Project Manager* for drug development projects: it is the role to do with the responsibility for managing a project. We then know that there must be an instance of the *Project Manager* role for the Xanthropol project, another instance for the Bisintifil project, and yet another for the Viniliton project. Each project has a separate responsibility associated with it.

When the Xanthropol project started, the responsibility for managing it was created. Suppose we appointed Bill to be the project manager. In Riva terminology 'We assigned Bill *as actor* of the instance of *Project Manager* for the Xanthropol project.' We might even say 'We cast Bill as actor of the instance of *Project Manager* for the Xanthropol project.'

Perhaps Bill then resigned, and Jill took over. The instance remained constant but the actor changed. Jill stepped into Bill's shoes; she took over the responsibility. But perhaps, just for a day after Bill's resignation, the project had no project manager. In other words the Xanthropol instance of the role *Project Manager* had no actor that day, no one to take responsibility. The important point to note here is that as long as a project exists, the responsibility for managing it also exists, *whether or not anyone is carrying it out, or acting it*. In other words, the role instance exists whether or not it has an actor.

So, we need to be careful to distinguish between the role instance – which has its own existence – and its current actor, if any.

Let's take a further example. *Prime Minister* is a role type. I can point to several instances of this role around the world: 'Prime Minister of the UK', 'Prime Minister of Australia' and 'Prime Minister of Canada', to name three. These role instances exist independently of whomever is acting them at any one time. As I write, a man called Tony Blair is acting the instance 'Prime Minister of the UK'. This instance was formerly acted by John Major. In fact Tony Blair acts a number of other role instances concurrently, one being that of 'Member of Parliament for Sedgefield', itself an instance of the role type *Member of Parliament*. In the Riva view of the world, an election is simply a way of choosing an actor for a role instance. If an MP dies, the role instance they were acting continues but there is no one to act it, no one to carry out the responsibilities of being MP for the area, until there has been an election.

If you approached an employee going into the publisher's building and asked them what they did, they might say 'I'm a Copy Editor'. They are saying 'I wear the hat of Copy Editor'. At home they do not wear that hat. (Let's assume they don't take work home.) So, as they walk into the building they don that hat and start carrying out the responsibility it implies.

Quite clearly, the role instance is separate from the person who acts it. That employee we stopped: when they go into a shop they don the hat of Customer. When they leave, they take the hat off. When they re-enter the publisher's building, they don the hat of Copy Editor again. And while they are wearing that hat they take part in various processes in which they have a part – a responsibility – to play.

Note how strong the theatrical analogy is: actors play roles. You might say 'I play the part of *CEO* in this company' in the same way that someone else might say they played the part of Hamlet in a recent performance.

Now an actor can take many forms. It might be a single person, a group of people, a computer, a person or group assisted by computers, a machine tool, a company – indeed, any agent in the real world capable of carrying out the work in the role.

When we model a process, we might (or might not) be concerned to understand where role instances come from and how actors get allocated. For instance if we are modelling how a new drug development project gets underway, we will at some point model how the project comes into being ('is born') and with that will come the creation of the responsibility for the management of that project: an instance of the role *Project Manager* will be created. We may also model, as part of the process, how a person is chosen to act that role instance – how they are 'cast' for the role; indeed, we shall allocate the responsibility of choosing someone to another role in the process, perhaps a *Therapeutic Area Director*. And we may go further and model how a replacement project manager is found if an existing one leaves the project.

A word of warning is in order here. Our first inclination when choosing roles might be to use names like *Store Manager*, but this can make it too seductive to identify roles solely with post or job titles, forgetting that job titles are invariably bundles of responsibilities in different roles in a number of different processes. For instance, the post of CEO could be treated as a role, but it is clearly a post that has a part to play in many processes in the company's activity: as authorizer of large purchase orders, as setter of the company's strategy, and as an important player in maintaining the company's relationship with its clients. Put another way, the CEO has a number of responsibilities in the organization. If we show the role CEO on the process model for, say, the process of setting company strategy, we must remember therefore that we are only saying 'This is the responsibility that the post of CEO has in this process' or 'This is the role of the CEO in this process'; we are not saying 'This is the CEO role.'

Equally, it is all too tempting to identify parts of the organization as roles: departments, divisions, sections or whatever. For instance, a Finance Department, though forming a readily identifiable group of people, might actually participate in a number of separate though related processes including remunerating staff (by paying people), purchasing (by placing orders and paying suppliers), and handling the company's cash flow (by invoicing clients, chasing bad debts and negotiating with the bank). It has a host of responsibilities.

Whether or not we associate posts or job titles or parts of the organization with roles will, like so many similar decisions, depend on why we are modelling the process, and later in the book we shall look at the various situations in which different approaches are appropriate. Some process models that I have prepared for clients have had, on a single page, roles from each of the different forms.

> **KEY POINTS**
> The *role type* defines the role.
> A role type can have a number of *instances.*
> Role instances operate independently and concurrently.
> Each instance can be *acted* by one or more people or no one, at a given moment in time.
> If appropriate, we can model the creation of new role instances.
> If appropriate, we can model the casting of actors in role instances.

The dynamics of roles, role instances and actors

Let's explore the relationship between roles, role instances and actors a little further.

In the case of Hill Pharm projects, we would expect to see only one instance of the role *Project Manager* being acted on a given drug development project, i.e. within one 'running' of the process that we might call **Develop a New Drug**. This is a situation where at most one instance of the role will exist when a process runs. Across the company there would of course be a number of instances of the role, one per active project.

The single instance of *Project Manager* could be acted by different people at different times during the lifetime of its project: Jack might be acting as project manager until 19 March, when he hands over the role (instance) to Jill. In a well-organized company we would not expect to discover that at some moment both Jack *and* Jill were acting the role, or a moment when nobody was acting it. So here is a case where we expect to find a one-to-one mapping between role instance and actor at any one time, though that mapping might change over time.

Let's think about a software product company which has a number of development projects in progress at any one time. Each project is responsible for the development and bringing to market of one new software product. Let's take one such project. Within that project, there could be several instances of the role of *Designer*, each being acted by a designer-type person, and, in a large project, perhaps hundreds of instances of *Software Programmer*, each being acted by a software programmer-type person. So when the **Develop a New Software Product** process is running for a particular software product, we would find one *Project Manager* instance, a number of *Designer* instances, and many *Software Programmer* instances. When we draw the model of the **Develop a New Software Product** process, we shall only describe the *role* types and hence imply that all programmers, for instance, follow the same procedure. If, in the real-life process, some programmers follow one procedure and others follow another, we would expect to see two different role types in our process model. For instance, novice programmers might be constrained by a rather detailed procedure to ensure that they do not do anything that threatens the project (and incidentally to train them in good practice), whilst experienced programmers might follow a less rigorously defined procedure which recognizes their greater expertise and grants them greater discretion. Our process model might therefore contain two role types that have the same goal but operate differently: *Experienced Software Programmer* and *Novice Software Programmer.*

In our **Develop a New Software Product** process, we might well have somewhere a role with responsibility for controlling change to specifications, designs, code, and so on: let us call it the *Change Controller* role. The *Change Controller* role will probably only have one instance on a project and that single instance might be acted by a whole team of people. Conversely, one (physical) actor might be acting several role instances simultaneously. For instance, on a small project, one person, Jill, might be acting the single instance of *Project Manager*, the single instance of *Change Controller*, and perhaps one of two instances of *Programmer*. Here is a case where the mapping from actors to role instances might be one-to-many.

Let's now look at how those role instances may come and go, and to do this let's revisit the different forms that a role can take, and look at how the number of instances and actors can change over time for each:

- A unique functional group: e.g. *Document Registry, Accounts.*
  Within a given organization we would expect to find only one instance of a (named) functional group such as these. In other words the role type (e.g. *Accounts*) has only one instance and that instance is effectively permanent; that is, we can take it as permanent for the purposes of the process model in which the role appears. The actors of the single role instance are of course the people (and perhaps computers) who make up the actual Accounts Department, and while the instance has a permanent existence (with the same qualification as above), the people who act it will change. Today I might be a member of Accounts, helping act the role; tomorrow I might have resigned and you might have taken my place.
  Suppose the organization we are concerned with is a nation. Then *The Government* will be a role, it will have one instance, and the actors will be those currently in power. Moreover, unless anarchy breaks out, the single instance is there for all time, with a succession of different actors acting it.
  Summarizing, there will be a single permanent instance of such a role with variable actor(s).

- A unique functional position or post: e.g. *Head of Analysis Department, CEO*.

  The situation here is similar to the unique functional group case: there is only one instance of the type. For instance there is only one Analysis Department and it has only one Head of Department post; similarly there is only one CEO post in this organization. The holder (actor) of such a post can change of course, but the post (role instance) is probably permanent for the purpose of the model.

  Summarizing, there will be a single permanent instance of such a role, with variable actor(s).

- A generic functional group: e.g. *Department, Branch*.

  Such a group will appear in a process model when we want to refer to *any* Department, Branch etc. and do not want to be specific about *which* Department we are concerned with, or we want to allow *any* department to play a particular part in the process. When the process runs, there may be any number of instances of the role – the branch in Oxford, the branch in Chicago, the branch in Auckland etc. These instances will be permanent for the duration of the process, unless of course the process is about the creation and closure of branches. The actors of this sort of role can of course change: staff at a given branch come and go.

- A generic functional position or post: e.g. *Head of Department, Divisional Manager*.

  The situation here is slightly different. At any one time there will, according to the organizational structure, be a fixed number of such posts: Head of Research Department, Head of Marketing Department, Head of Production Department, and so on. We might want to consider each of these as an instance of the role *Head of Department*. And each role instance – each Department Head post – will have an actor: the current holder of that post.

  Summarizing, in general the set of instances will remain fixed for a given process, but there will be a change in actors as people are put in those posts and leave them.

- A generic type of person: e.g. *Trade Union Member, Customer, Purchaser, Expense Claimant*.

  This is like the generic functional group: when a process runs, one or more instances will be created but each will be identifiable with an individual, and the role instance is in a one-to-one relationship with that individual.

  If an instance of *Customer* comes into being it will be associated with and acted by a single person; the actor will not change. During the handling of a complaint from Mr Bloggs we do not expect to see Mrs Featherstonehaugh take over at some point: the *Complainant* instance's actor remains constant.

In the process **Sell a House**, we shall have a role *Vendor*, representing the seller of the house concerned. For a given house there is one vendor, so the role will have at most one instance during the process. And the actor of that instance, the person or group selling the house, is very unlikely to change during the sale.

On the other hand, in the process **Find and Buy a House**, the role *Vendor* will represent the sellers of houses we might buy. Instances of *Vendor* will come and go, as houses come onto the market and come off it during the process. Each instance will be acted by a fixed actor: the owner of the house concerned.

The *Expense Claimant* role is the hat we put on when we want to claim expenses; it is the role we act to claim expenses. Our main role might be *Supervisor*, but once a month we 'slip into' the role of *Expense Claimant*.

- An abstraction: e.g. *Progress Chasing, Project Managing*.
  This is a case where a gerund (an -ing noun) makes a good name for the role. It can be particularly useful to name an area of responsibility, rather than a box on the organogram for the organization, such as a post or a department. I might start *Progress Chasing* an invoice but someone else might take over at some point if I ask them to; and there could be any number of us progress-chasing various items at any one time. Be careful not to think of this as an action: it is a responsibility within which certain actions will be carried out, but we have yet to look inside the box and examine actions.

One apparent flavour of role that we need to be careful with is a *rank* or *job title*: e.g. *Principal Analyst, Senior Engineer Grade 5*. This is a 'badge' that people have, that they carry round the organization, and that brings responsibilities or entitles them to carry out certain responsibilities. When we label a role in a process with such a badge we are saying 'Anyone acting this role – carrying out these responsibilities in the process – must have this badge.'

KEY POINTS

A role instance generally exists independently of an actor to act it.
Generally, the actor of a role instance can change.
At a given moment, there might be no one acting a given role instance.

Role instances start other role instances

We have seen how some roles have *permanent* instances, permanent in that they are there when the process concerned starts and still there when it finishes. (Remember we are looking at a single process at the moment.) Suppose we are modelling the **Prepare the Annual Budget** process. We might expect that the unique role *CEO* plays a part. When the process of preparing the budget starts, that instance already exists (moreover, when the process is over and the budget is ready, the instance is still there). To be precise, we shall say that 'The role *CEO* has a pre-existing instance for the process.'

However some roles may not have instances when the process starts. Suppose we are modelling a process called **Develop a Software Product**. We could imagine that during the early phases of the process/project, we would want to create a role instance to manage change requests and appoint someone to carry out that role instance. Once the instance is in place and has an actor assigned, the management of change requests can begin. Clearly in this situation we will need instances to be created during the process, and it will not come as a surprise that one role instance can cause the creation of new role instances to create new responsibilities. This corresponds to what happens in real life: one responsibility can create another responsibility. We shall say that one role can 'instantiate' another. For example, in **Develop a Software Product**, the (single) instance of the *Project Manager* role might instantiate the *Change Manager* role. This is what happens when a Project Manager creates the responsibility for managing change.

In the pharmaceutical R and D industry, when a chemical compound looks as though it has promising therapeutic properties, a team is often set up to champion that compound through the long process of getting it to market. That team has responsibility for managing the compound's life thereafter. So the *Compound Management Team* role is instantiated to carry out that responsibility. During the compound's development, many batches of raw drug material will be made for formulation for clinical trials. The (abstract) role of *Making a Batch* is instantiated for each batch – each time a batch is to be made there is a responsibility created for that; in fact, there may be any number of instances of that role at any given time, depending on how many batches are in the process of being made at that time. (Remember that Making a Batch is not the name of an activity: it is the name of the responsibility for making a batch.)

Any role that is not permanent as far as the process is concerned will have to be instantiated at some point – we have called these transient roles.

Note that creating a new role instance does not imply any allocation of real people or machines to act the role instance, as far as a RAD is concerned. The *actual* team of people who manage the life of a candidate compound may well change over the many years during which the compound is developed, but the role instance remains. When a batch of compound is to be made, the role instance is started and a process research chemist is allocated to actually make the batch, i.e. to act the role instance – it could be John or it could be Jill. Again we are being careful to separate a role instance from its actor. Also, if the new role instance needs resources (props) to carry out its responsibility, we might be interested in modelling how it acquires them. When I start work as the manager of a project, what props can I expect to find on my desk, where did they come from, and which props do I have to acquire for myself?

KEY POINTS

Some role types have pre-existing instances in a given process.
Other role types must be instantiated during the process: this equates to creating a new responsibility during a process.
Instantiating a role type only creates the responsibility; allocating actors to carry out that responsibility is a separate matter.
A role type has an *instance profile* in a given process:

- It can have one or more pre-existing instances.

- Those instances might be fixed for the duration of the process.

- The actor of a role instance might change during the life of the instance.

That 'in a given process' above is important: a role's instance profile will depend on the scope of the process context. As far as the process of taking a proposed law through Parliament is concerned, the role *Member of Parliament* can be considered to have 635 pre-existing instances whose actors are of no concern to us. But if we are concerned with the process of electing new Members of Parliament, then the actors and their allocation is very much a matter of concern. If we are concerned with the (unusual) process of redefining parliamentary boundaries, and hence with creating and deleting instances of *Member of Parliament*, then we shall have a certain number of pre-existing instances, and potentially a different number of instances on completion of the process; actors will not be our concern here.

Later, next to the water cooler

Pupil:	Something puzzles me. You're saying – and I understand this – that responsibilities come and go in the organization. To take the example earlier, when a new project comes along in Hill Pharm, that creates a new responsibility for managing the project. Once the project is finished, that responsibility is gone. So there is a flux of responsibilities washing through the organization – right?
Tutor:	Right. In fact I like the word *flux* – a continuous succession of changes ... that's what's happening: responsibilities are coming and going.
Pupil:	But the organogram stays more or less constant. In particular, the organizational structure doesn't change with the flow of responsibilities. What's happening?
Tutor:	Simple. Every time a new responsibility arises it gets allocated to a post or a functional group, or one of the other concrete – let us say 'real-life' – roles. When a purchase order is raised, the responsibility for checking and approving it is created. That responsibility is an abstract role. I would call it *Purchase Order Approving*. As you say, as time passes there is a flux of such role instances, coming and going. Yet all invoices are actually handed to the

	CFO for checking and approval. So we are allocating all the *Purchase Order Approving* role instances onto the single instance of CFO. Try the idea with a different situation: suppose someone calls the Helpdesk.
Pupil:	OK, a call comes into the Helpdesk. That creates a responsibility … to answer it satisfactorily, say. I guess you'd want to call that role *Call Handling*. One way or another the call is routed to a desk. I guess the desk represents a concrete role. But you would want me to be more accurate: the desk represents an instance of a concrete role – call it *Helpdesk Station* – and the person sitting at that desk is acting that instance. So we have mapped that instance of *Call Handling* onto that instance of *Helpdesk Station*.
Tutor:	Exactly! Now do the same with another situation: the fire service gets a report of a fire.
Pupil:	Right, well, the report creates a responsibility – to put out the fire. The responsibility is taken on by the fire crew leader. So the responsibility, which we might give the abstract role name *Fire Handling*, has been mapped onto the concrete role *Fire Crew Leader*. Presumably, when we model all these processes, we have a choice between modelling the abstract role – a transient role that comes and goes – or the concrete role that the responsibility is mapped onto, which is rather more permanent?
Tutor:	Precisely. Let me try and capture this. We identified six types of thing we might find 'playing a part' in a process: unique post, unique functional group etc. Five of them are organizational realities, the sixth is an abstraction: a 'pure responsibility'. In many situations we create responsibilities on-the-fly. We'll see later that such a responsibility typically 'goes with' what we shall call a unit of work (UOW); it's the responsibility for that UOW. But that responsibility has to be allocated to something in the organization: one of the five organizational types of role.
Pupil:	So when a customer order arrives we might say that the responsibility to deal with it is created, and that we give that responsibility (initially at least) to, say, a *Customer Order Clerk*. On the RAD, we could choose to model the abstract role – the responsibility – *Customer Order Handling*, or the organizational entity *Customer Order Clerk*.
Tutor:	A good example. A post is another example. The organizational post CEO means nothing on its own. It is given meaning by the responsibilities that we allocate to it. The post CEO takes on many responsibilities in many processes. Some are permanent, many are transient. By acting as *CEO*, a person takes on that flux of responsibilities.
Pupil:	One final point: organizations depend a great deal these days on their computer systems. They have systems, often massive systems, that – shall I say – play a part in the process. Would you consider such a system as a role?
Tutor:	In a sense, yes, but I'm going to be a bit cautious about the idea. Later on we shall indeed see some process models that have such systems as roles, playing a part in the process. I'm cautious simply because it's hard to say that the system 'takes responsibility' for anything – it is just a heap of metal and plastic after all. You're looking tired?

Pupil:	You really have whittled away at this idea of a role. Why should we get involved with all these subtleties?
Tutor:	Have patience. We shall have some important questions to answer and these categories of role will help us to answer them:

- Do we understand the dynamics of each role?

- Do we need to model how instances arise? In particular, how do responsibilities get identified and created and by whom? Who has the authority to create a new responsibility in the organization?

- Do we need to model how actors become connected to the instances they are acting? The relationship between actor and role instance is what the scheduling of staff and resources is all about, and this itself might be part of the process or indeed another process altogether.

Pupil:	OK. I can see how it would be all too easy when looking at processes to worry just about what is done, and to forget how things get organized to get done, and how responsibilities are created and handed around.
Tutor:	Indeed, but there's more to it than that. A constant theme of Riva is that organizational activity is massively concurrent: there are many, many things going on at the same time. If we want to get our heads round it and get a true and full understanding of that organizational activity, one that captures that concurrency, then we'll need ways of describing it in the language we use. Role instances are one of the sorts of concurrency that we find. Role instances have their own lives: they operate concurrently. We'll see later in this chapter that they also operate independently except where they collaborate through interactions. From now on, we must take care to differentiate between a *role type*, a *role instance*, and the *actor* of a role instance, unless the sense is clear from the context.

(If you are familiar with the notion of *use cases* in the Unified Modeling Language (UML), you will need to make a concept switch from UML's use of *role* and *actor* to Riva's. UML confusingly defines a *use case* as '*The specification of* [my italics] a sequence of actions ... that a system (or other entity) can perform, interacting with actors of the system', an *actor* as 'A coherent set of roles that users of use cases play when interacting with these use cases'. An actor has one role for each use case with which it communicates, and a *role* as 'The named specific behaviour of an entity participating in a particular context'. (See www.omg.org/uml.)

ACTIONS

When we look at a single process, we know that we chunk everything that gets done in the process into roles, and that each role represents an area of responsibility within the process. We must now look inside each role.

Actions are what actors do on their own in their roles to carry out their responsibilities. Let's take some examples:

- In the process of developing a new pharmaceutical drug, actions might include *Choose lead molecule, Carry out a clinical trial, Prepare the submission to Regulatory Authority* and *Clean the pilot plant.*

- In the process of buying a house we might find actions such as *Choose estate agents (realtors), Obtain finance, Obtain planning permission* and *Negotiate the price.*

- In the process of developing a software system, actions might include *Prepare the project plan, Prepare use case model, Carry out a proof obligation, Transform an algorithm, Verify a code module against specification, Build the system,* and *Add a component to the object library.*

Note how we name actions with verbs: 'prepare', 'draw up', 'verify' etc.

An action needs to be well defined; in particular we need to know what makes it start and what makes it stop, what state the world is in when it starts and what state the world is in when it stops, i.e. when and why an action is done. Let's look at this in more detail.

Like roles, actions are defined in Riva as types that have instances. If we are speaking informally about a process we shall say 'Action *A* starts'; if we want to be formal and precise we shall say 'An instance of action type *A* is created.' An instance of an action type is created when the organization's process enters a particular state: we call this the *activating condition* for the action. This is a sufficient condition for the action instance to start. For example, in an organization that has a policy of paying invoices three months after receiving them, the action *Pay invoiced amount* would have as its activating condition the fact that payment of an invoice has been due for three months. In an organization that pays on receipt, the activating condition would be the fact that an invoice has been received. The *post-condition* of an action is the *state* of the world when the action (instance) has finished. *Pay invoiced amount* would probably have as (part of) its post-condition the fact that a cheque for the invoiced amount had been sent to the supplier.

Not surprisingly, the post-condition of one action will often be the activating condition for another. So in the **Purchase Materials** process in our prompt-paying organization, the post-condition of the action *Receive invoice* – probably *Invoice received* – will be the activating condition of the action *Pay invoiced amount.* In this case we might say that 'the action *Pay invoiced amount* follows the action *Receive invoice*' or that 'the action *Pay invoiced amount* is consequent on the action *Receive invoice*'.

There may be other conditions that are also true when an action is started and which we want to note (though they are not what makes the action start); these are collectively referred to as the action's *pre-condition.* They are necessary but not sufficient conditions for the action to start. Thus, in the late-paying organization, the action *Pay invoiced amount* might have as a pre-condition the fact that payment has been authorized.

Authorization does not activate payment: 'due for three months' does. Payment cannot occur unless there has been authorization.

(Strictly, a post-condition is a necessary but not sufficient condition for completion of the action. We can also define a *stopping condition* of an action, which is the sufficient condition for the action instance to stop.)

As an example in **Develop a Software System**, suppose we have an action (type) called *Compile component source code*. The activating condition for this could be *Component source code successfully syntax-checked*; a pre-condition could be *Access available to validated library*; the post-condition could be *Corresponding object file available in the object directory*.

Finally, let's note that an action can have alternative activating conditions: there can be a number of different situations that cause me to send an email, or write out a cheque, or compile some source code. Each has a different activating condition, but they will all share the same pre-conditions.

Actions change the state of things

Pupil:	I've noticed that you've made no attempt to define an action in terms of *inputs* and *outputs*. I've seen actions described as ways of '*transforming inputs into outputs*'. Why don't we do that in Riva?
Tutor:	Because that manufacturing-oriented way of thinking about actions and processes isn't helpful. We end up distorting the idea of an input and an output – and hence our understanding of an action – just to stick with the metaphor. The compiled object code could certainly be thought of as an 'output' of the action *Compile component source code*, a product of the action. But it's nonsense to say that the source code was transformed into the object code. It clearly wasn't. The source code of a software component isn't *consumed* by the action – it still exists after the compilation – so calling it an input is strange. A dirty car is not consumed by the act of washing it.
Pupil:	No, I guess not. The purpose of washing a car is to change its state from 'dirty' to 'clean'. Ah … I described it in terms of states. But couldn't we say that a dirty car was the input and a clean car was the output?
Tutor:	We could, and one way I could implement that is to crush and dispose of the dirty car you give me and hand you back a clean car. I didn't transform the car you gave me but I satisfied the input–output specification. What sense does it make to say that a purchase order is transformed into goods? The whole input–transform–output metaphor leads us into absurd statements and is best avoided.
	It is much more natural and less forced to think of the *state* of the world before an action starts – there is a particular software component that is uncompiled – and its state after the action has finished – that there is an object-code file associated with the component source code in the development library. The purpose of the action is to change the state of (that part of) the world. This all becomes much more obvious when we are dealing with desired outcomes such as 'a happy customer'. I would rather

say that an action leaves a customer in a happy state, rather than that it outputs a happy customer – as though they have popped out of the side of the box smiling, having gone in the other side looking glum. And I would certainly not want to say that an action 'transforms a purchase order into a happy customer'. That sounds like science fiction.

Pupil: One example that comes to mind is the process for curing someone of an illness. The input – forgive me – is a sick patient and the output is a well patient. What's wrong with that?

Tutor: Remember my constant message: when we think about processes, we are thinking about *dynamics*. When we think about actions, we want to know when things happen, *when* actions start. States do that for us. When we say '*C* is the activating condition for action *A*' we are defining the dynamics of that part of the process: *C* is what makes *A* start. Defining the inputs of *A* (whatever it means) doesn't tell us what makes *A* start. Inputs and outputs might capture data dynamics but they don't capture process dynamics. States do.

If you tell me that the action *Cure a sick person* has a sick person as its input and a well person as its output, I'm no wiser about what starts the activity, which seems to me to be vital information. I need to know that its activating state is *Sick person waiting in reception*, and that its post-condition is *Well person has returned home*, or possibly – let's be realistic – *Dead person is in mortuary*. (Of course, if we do allow that other possible post-condition, we have named the action badly.)

Finally, we shall see in a moment the importance of *goals* of a process. Goals are, of course, just desired states, so we shall need to have the language in place for talking about states.

Actions have relationships

Actions not only have important properties of their own, they also relate to each other in different ways. There are three ways:

- Action *A* might always follow action *B* in role *R*: a cheque cannot be sent until the expense claim has been approved. So actions may be *ordered* and follow a particular sequence. The action *Approve expense claim* must precede the action *Send cheque for expenses* within the role *Finance*.

- Either action *A* is carried out or action *B* is carried out in role *R* depending on whether some condition *C* holds: if the expense claim is over £1,000 it is paid by electronic transfer, otherwise by cheque. So actions may be *conditional*. The action *Pay expense claim by electronic transfer* can only start (be instantiated) if the condition *expense claim exceeds £1,000* is true; the action *Pay expense claim by cheque* can only start (be instantiated) if it is false.

- At some point both action *A* and action *B* can proceed in parallel in role *R*: once the expense claim has been approved, the money can be paid to the claimant and the relevant department budget can be debited. So actions may be *concurrent*. The action *Pay expenses to claimant* can proceed concurrently with *Debit departmental budget* within the role *Finance*.

In our process model we will want to be able to show where actions are sequential, conditional or concurrent. These will be one way in which business rules will be represented.

KEY POINTS

An action is carried out by a role on its own.
The activating condition of an action is a state that causes an action to start (be instantiated).
The post-condition of an action is the state of the world when the action has finished.
We define the dynamics of roles in terms of state changes.

INTERACTIONS

Roles 'chunk' the activity in a process. Roles carry out actions on their own account. But we started from the important axiom that the main way things happen in a process, especially in terms of moving the process on or making progress, is through the *interactions* that take place between roles, such as when a manager delegates a task to a subordinate or a price is negotiated.

In the process **Develop a Software System for a Client**, the role *Project Manager* will want to interact with the roles *Designer* and *Programmer* to obtain status reports on work completed, and the *Designer* role will want to pass specifications of programs to be written to the *Programmer* role. In the process **Develop a Portfolio of Products**, the *Board of Directors* will want to pass a statement of direction to the *Product Strategy Board* along with a budget level and targets. In return the *Product Strategy Board* will present the *Board of Directors* with information on the chosen portfolio, and progress reports against budget and targets.

In Riva terms, an interaction is neutral and has no implied direction – it is just some *coordination* between roles, a collaborative act. But an interaction might involve the transfer of something – what we shall call a *gram* – from the body of one role to that of the other. For example, the *Divisional Manager* role interacts with the *Project Manager* role so that the former can pass the latter some terms of reference for the project they are to manage. But an interaction need not involve the transfer of a gram: for instance, you and I might interact simply to agree on something – 'nothing changes hands'. For instance, the Sales Team, the Marketing Team, and the Production Group of a product company might collectively decide when a new product should go to market: that interaction might consist of a discussion around a table.

As with actions, we will think of interactions in terms of states and state changes, not in terms of inputs and outputs. When I reimburse you with your expenses, I see it in terms of your change of state: before the interaction you didn't have your expenses, after the interaction you did; before the interaction I had the money, afterwards you had it.

An interaction can be two-party – involving two role instances – or multi-party, involving a number of role instances. However many parties are involved, an interaction is always *synchronous*:

- it starts at the same moment for each party, as soon as they are all ready;
- it completes at the same moment for each party, as soon as they have all finished.

One way of thinking about an interaction is as *an alignment of states*. For you and me to interact, we must both be 'waiting' in the required state beforehand; we go through the interaction together, and then we go to our respective after-states.

In some cases, an interaction might physically take a few seconds (I give you a memo containing some terms of reference), in others, months (a vendor and purchaser agree on the contractual terms of a sale). As elsewhere in our modelling notation, we do not capture absolute time: there is no time axis on our models. We might choose to annotate actions and interactions with their duration in some way, but no more.

> **KEY POINTS**
> Interactions between roles are the way that collaboration happens in a process.
> Interactions are the way that role instances coordinate their activity.
> In some interactions grams change hands.
> Interactions align the states of the interacting parties.

PROCESS GOALS

Point-wise goals

Processes are there for a reason. For instance, the goal of a process might be to deliver a computer system, to provide a medical procedure to a patient, or to manage a research budget. It must be possible to see from our process models how a process is achieving the goals set for it, and ideally, to be able to identify the point(s) in the process where those goals can be said to have been achieved or maintained.

In the simplest case, we will be able to identify some point in the activity of a particular role where the state of the process is 'goal achieved'. After a particular action or interaction has completed we can recognize that the goal has been achieved. For instance, in a process for handling a reported credit card loss, we might say that once there has been an interaction with the customer in which the latter has been sent a new credit card, the goal *Client has been sent replacement credit card* has been achieved. We can identify the role in which, at some point, that state has been achieved, i.e. the state at that point *is* the goal. Reaching that state is reaching the goal.

The goal of an insurance company in its New Policy Applications Department might be to respond to a customer with a proposal within seven days of receipt of their application. Each step in the process can contribute to the successful achievement of that 'point-wise' goal: there comes a point when we can say 'Here the goal of the process has been achieved.' The goal of the process of carrying out a software development project for a client is to satisfy the client with a timely delivery of working software. The only point at which we can check whether we have been successful is the point of delivery: is it on time and does it work? Until that point we can only make predictions.

In some situations there may be several goals, leading not to one state in one role but a combination of states in a number of roles. For instance, the goals of the **Handle a Reported Credit Card Loss** process might be not only that a new card is sent to the customer, but also that the credit card fraud bureau is informed and that the card number is entered on the list of lost cards that is circulated to retailers.

The New Policy Applications Department in a life insurance company wants to get a correct policy proposal out within seven days of receipt of the application; it might also want to ensure that the business it takes on is good business, i.e. that the premiums it charges adequately cover the risk; and it might also have the goal of making the terms offered available to those assessing the competitiveness of the company's products.

On the way towards achieving a final goal, we can often identify *sub-goals* which represent milestones of some sort. Issuing the insurance proposal to the customer requires that, at some point en route, the risk be satisfactorily assessed and that approval be obtained for the premium offered.

This indicates, I hope, how important it is to see a process model as a description of *the way that the organization changes state*, from its initial state (someone wanting a service or whatever) to its final state (service delivered).

Steady-state goals

When I run my life in financial terms, one of my aims is to regulate my earning and spending so as to keep my bank balance in credit. At any moment in time, I prefer to have the bank looking after my money rather than the other way round. I don't like paying interest. The critical bit is 'at any moment in time'. I want to see a steady state in which I am in credit.

The steady-state goal is more complicated. By definition, it says that something is true at all times or – softening this a little – that it is true at a number of points in the process (the times when we choose to look, say). I might want to be sure that at all times everyone has the latest information on product features, or that expenditure is always kept within the budget level set at the start of the year. The process of managing cash flow in a company has the same goal: that of maintaining a steady state in which the company keeps a positive cash flow as expenditure and revenue rise and fall. If the process is successful we should be able to observe the company at any moment in time and note a positive cash flow. We shall unpick this a little more later.

KEY POINTS

Processes have goals.
Goals are simply desired states.
Goals can be point-wise or steady state.

ENTITIES

Pupil:	Why isn't *data* or *information* in our list of concepts?
Tutor:	The simple answer is that in this approach to business processes we concentrate unashamedly on what people *do*, rather than on what people do it to or what they do it with. Once we have chosen a structure for the organization and the processes it will operate, then we can decide on the information needed by individuals and groups to perform those processes in that organizational structure.
	Process precedes information.
	Once we have a process model – a description of how the business does its business or plans to do its business – we can start to investigate the information needs of the process: Who needs what information to do that or to make that decision? And how does that information get to that person? We might think of documents, in particular, as the oil that lets the wheels turn, but they aren't the wheels!
Pupil:	You've made a lot of disapproving noises about inputs and outputs. And now you're keeping low on information. Do you deny the existence of *things* altogether? For instance, a manager will prepare a Plan, write a Report, draw up Terms of Reference, or approve a Document. Plan, Report, Terms of Reference, and Document are entities. A programmer uses a Specification during the programming action and produces a Program. A warehouse person uses a Picking List during the packing action.
Tutor:	Yes, we shall indeed talk about *entities*. And we'll use the word 'entity' for anything that is the subject matter of an action. Or indeed an interaction: when I delegate a task to you I perhaps pass you some Terms of Reference; and when you have finished the work you will pass me the Results. The grams of an interaction are frequently entities.

Like roles and actions, Riva entities come as *types* which can be instantiated. In fact, if an action instance produces an output we consider it to be *instantiating* the relevant entity type.

When we define an entity, we will want to ascribe properties to it. Amongst those that are covered by Riva are the parts from which a compound entity is composed if it is a composite thing and *invariants*, i.e. things that are always true about the entity. For example, a Technical Specification might be defined to be made up of a Contents List, Document Control Section, a Scope Section, a Control Flow Section, and a Data Flow Section, followed by Performance Details. A Production Plan might be made up of a List of Input Resources, a Timetable, and a Definition of the process to be used. An invariant of the entity Production Lathe is that it is *up to date in its maintenance schedule.*

Pupil: And entities do play a part, don't they? If I ring up someone who is doing something for me and ask them how things are getting on, they'll often tell me about the state of the objects that are involved: 'Your application forms have been approved', 'Your car will be ready in half an hour', 'The catalogue you ordered is on its way to you', and so on.

Tutor: That's right. And more generally, when we define the activating condition, pre-conditions or post-condition of an action, we will often do it in terms of the states of entities. For example:

- The activating condition of the action *Send Invoice* is that there is an Invoice and it has been approved.

- A pre-condition of the action *Send Invoice* is that the goods have been received in good condition.

- A post-condition of the action Send Invoice is that the Database has been updated with the date the invoice was sent, that the invoice has been sent, and so on.

Pupil: So we will have entities, but information – presumably about those entities – is simply their state? And we shall talk about state? And we shan't concentrate on information per se?

Tutor: Exactly. If our task is to understand a process, then worrying about the information of the process is a bit like trying to understand a document by worrying about the typeface in which it is written.

KEY POINTS

Entities are the subject matter of actions and interactions.
Entities are instantiated, just like any other types.
Entities have states, some of which are used in defining actions and interactions.
Information consists of state descriptions.

THINGS ARE COMPLICATED

Pupil: You've outlined the ideas behind each of the central concepts – roles, actions, interactions, and so on – but I sense that there is a more general message that you're getting at. This notion of *instance* seems important.

Tutor: Yes, in fact there are two important 'inner' messages here. The first is the central place of *instantiation*: the making of new instances. It's important because it's what drives the *dynamics* of a process. We've seen how roles in particular are instantiated: new responsibilities are created dynamically during a process. We'll look more closely later at the instantiation of actions and interactions.

Pupil: OK, so an organization at any one moment is just a mass of instances, and an organization viewed over time is a flux of instances. What's the second inner message?

Tutor: It's a closely related message: processes are about concurrency: lots of things are happening at the same time. When we walk into the building and watch the mass of organizational activity, we don't see a simple flowchart being followed, with someone's finger tracing down the boxes. We see many instances of many different processes in progress simultaneously. Within each process instance we see many role instances active at the same time. We see role instances coming and going as responsibilities arise and cease. And within each role instance we see potentially many threads of activity in progress. In summary, there is a massive network of interrelated threads operating.

Pupil: So, all those instances mean concurrency. And presumably, what an organization achieves it achieves because of all that concurrency – things don't happen because of sequential activity.

Tutor: Exactly. Let me give you an analogy which my colleague Clive Roberts embodies in the logo of his company Co-ordination Systems. When geese fly long distances, each bird uses the same tactic: fly in a specific relationship to the next bird. The result is a chevron of birds. That chevron is an *emergent behaviour* resulting from the interactions and concurrent activity of all the birds. Yet no bird looks at the overall shape and decides how to respond. It follows its own tactic. The effect of the organization is an emergent property resulting from the interactions and concurrent activity of many role instances across many process instances.

To truly understand our processes as dynamic objects we must grapple with instantiation and with concurrency, and in particular find ways to model them. And, if we want to design sets of processes that we can actually execute on a Business Process Management System of the sort we shall explore in Chapter 13, we shall need to be able to design those dynamics that give the right emergent behaviour.

Instantiation and concurrency are central themes of Riva, and when we have developed the argument we shall turn to the idea of *mobility* of processes.

2 Modelling a process

Provides all the vocabulary necessary to represent a single process in a Role Activity Diagram (RAD).

SOME HEALTH WARNINGS

We have now looked at our needs as process modellers and at the aspects of the real world that we shall want to see in our models if we are going to satisfy those needs. It's time to look at the language for our process models.

But I want to start with two warnings.

We are going to begin by modelling a single process. But this begs a rather important question: how did we decide that this process exists? Put another way, how do we know it makes sense to call this particular 'pile' of activity a 'process'? In any organization we might guess there will be many different processes. We might guess that they are related: that the process for purchasing goods is in some way related to the process for dealing with invoices; that the process for opening a new bank account for a customer is in some way related to the process for checking a person's credit; and so on. So how did we chunk all the activity in the organization into those processes and then choose this one to model? These are questions we must, tantalizingly, leave unanswered until Chapter 6. Once we have the language necessary to model a single process, we can look at the different sorts of relationships that can exist between processes, and then build up a theory of 'chunking' that will allow us to answer these crucial questions.

This introduces a danger: if you don't read further than this chapter you may go off and start modelling things that you think are processes but which could only be called 'collections of activities', collections that don't have the coherence that exists in the real world. The axe man cometh.

That's the first warning: *be sure that what you are modelling is a process.*

Here's the second warning: *there is no single model of a process.* Our viewpoint will vary as our motives vary. If we are interested in why a process seems to bottleneck in certain areas, we might want to model the process from the point of view of how work is allocated to individuals. If we are interested in how the functional subdivisions of the organization help or hinder the flow of a transaction through a process that crosses the functional boundaries, we might want to view the process in terms of those boundaries and the interactions across them, without worrying too much about *how* each function does its work.

There are as many models of a single process as there are viewpoints that we might want to take. The perspective taken, what is left in and what is left out – all these decisions rest on our judgement as modellers. Again no simple rule will say what perspective we should take in any given situation, though in later chapters we shall draw up some guidelines. This is a point that I shall return to many times in the book. Relevance is in the eye of the modeller, too. A model is 'right' . . . if it helps. It helps if it *reveals* things we want to know, or helps us answer questions, or can be analysed, or can be adjusted to test proposed changes, or simply aids understanding. We shall see in Chapter 8 that the most important thing at the start of a modelling activity is to be clear about the *purpose* of the model. The model itself cannot do anything – it is a tool that will work well in the right hands in the right situation, and badly otherwise.

Indeed, we might take several perspectives corresponding to the different perceived purposes of the process we are looking at. In our work for a pharmaceutical company, Tim Huckvale and I initially prepared two models of a particular process, essentially seeing the same process from two perspectives: one from that of the scientists doing the science necessary to take a new drug compound to market, another from that of the management pushing the development of the compound through the various stages of process scale-up and trials whilst weeding out those compounds that would not offer future success. In Soft Systems terms, these views could be considered to be *holons* which we 'put against the world' in order to learn about it (see for instance Checkland and Scholes, 1990, and Patching, 1990). Each corresponds to a different idea of the purpose of the process (/system). The Research Chemist saw the purpose of the process we were looking at to be to produce a way of making the drug compound safely in the required quantities and to the required purity; the Regulatory Affairs Group saw its purpose to be the production of the information and audit trail of development which would satisfy the industry regulators; the Clinician saw its purpose to be the timely production of sufficient quantities of drug for the clinical trials they were planning; the senior management saw the purpose to be either to get a successful-looking compound to manufacturing as quickly as possible or to drop an unsuccessful-looking compound as early as possible; and so on.

If I walk into a map store and ask for a map, I'll hope I'm asked 'What do you want the map for?' I could have a number of reasons:

- To walk from Paddington railway station to Victoria railway station in London. I need a map to help me find my way around. In particular, it will need to be a fairly detailed map as I am interested in being able to trace my steps through London's network of streets. I shall need street names but won't need to know if streets are one-way for vehicles.

- To drive from Bath to Birmingham. Again I'm looking for help in finding my way around, but now I shall need a map on a different scale: one that shows me the broad shape of the country and the major roads will do. I won't need anything showing country lanes, small villages, or the local topography of the countryside I shall be passing through. It would be useful if the map also had outline street maps of Bath and Birmingham.

- To allow me to agree with someone on a spot in London where we will meet. In this case my reason for needing a map involves someone else: we will use the map as an agreed definition of something; we can agree on where we will meet in predefined surroundings. If we are meeting at a pub, it would be good if the map showed the locations of pubs.

- To agree on a boundary to be drawn on the sale of some land. Here we want to define something not already defined, and to place it in some larger context, with references to existing features.

- To decide where to move an existing footpath. In England a footpath may well go back many centuries; its end points and its route will have been determined by needs from the past. Changes in surroundings might make moving it sensible. We need to know what the options are. We will be working at quite a small scale.

- To decide where to route a new road. If I am planning a completely new road from A to B, I shall need to explore the options and the impact that each will have on things that I want untouched. I want a map that shows topography on a large scale, at a level of detail that allows the exploration of impact.

- To record the position of underground cables. Here, as a cable company say, I shall be maintaining my own maps of where my cables run relative to the infrastructure of the town. I need these maps to allow my staff to find the cables in the future.

The parallels between maps and process models should be clear. Before we carry out some process modelling we will need to know quite clearly what we want from the model so that we can choose the scale it is at, and the sorts of detail it shows.

KEY POINTS

Before modelling a process, be sure it is a process – always start with a process architecture.

There is no such thing as *the* model of a process.

Worse, all models are wrong … but some are useful.

To get to useful answers we must ask the right questions.

Our choice of model is guided by the questions we want to answer.

Before you start, answer the question 'What questions do I want to answer with this model?' and write your answer on the wall.

THE ROLE ACTIVITY DIAGRAM

A Riva process model takes the form of a RAD. A RAD shows the roles that play a part in the process, and their component actions and interactions, together with external events and the logic that determines which actions are carried out when. So, it shows the activity of roles in the process and how they collaborate.

I have summarized the notation for RADs – the various sorts of blobs – in Figure 2.1. This is the language we shall use. In this chapter we shall look at each of the elements of this language and take each one in two steps: how to use it diagrammatically, and what to do in different modelling situations. (The little spring-shaped symbol means 'don't care'. It is a sort of pictorial ellipsis ... and you will find it used a lot in sample RADs throughout the book. If it appears at the start of a thread it means 'We don't care how we got here.' If it appears at the end of a thread it means 'We don't care what happens after here.' If it appears in the middle of a thread it means 'We don't care how we cross this gap.')

Figure 2.2 shows a RAD for a simple process, to give you a feel for what a complete RAD looks like.

A RAD represents the whole of a process as far as we wish to capture it. Somewhere on the RAD we name the process it is modelling. This might be **Develop a Software System for a Client, Develop a Portfolio of Products, Arrange a Payment of Benefits** or **Manage the flow of Customer Queries**. Remember, we haven't yet discussed just what constitutes a process, how to decide what goes in one process rather than another, or how processes fit together – all of this will come later.

Let's now look in detail at the RAD notation and how we use it to capture the concepts covered in Chapter 1.

(Computer tools that support the preparation of RADs may use slightly different symbols but the shape is of no importance. It is the *meaning* we attach to those symbols that is important, and it is that meaning that this chapter addresses.)

REPRESENTING ROLES

Each role in the process is represented by a shaded block with rounded edges. Everything the role does appears in that box and since only roles can do things, everything on a RAD is inside a role box. In Figure 2.2 there are three roles with the names *Divisional Director, Project Manager,* and *Designer.* All actions and interactions take place within those three roles. We can turn this around and say that everything is done as part of carrying out some responsibility or other.

The name of the role appears immediately above or below the block, whichever is convenient.

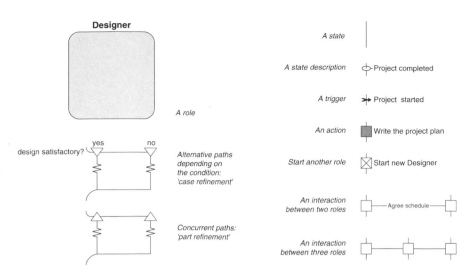

FIGURE 2.1 *The RAD notation*

We sometimes draw a single role as a number of separate shaded blocks if there are indeed separate parts of the role and it makes the RAD easier to draw. Sometimes, as another modelling convenience, it is handy to let the boxes of different roles overlap, provided of course there is no ambiguity about what is in which role. Where we do overlap them we use different shading to distinguish the two roles – see Figure 3.5 as an example of both of these modelling conveniences.

In Chapter 1 we examined several different sorts of role: unique functional group, type of person etc. Our notation does not differentiate these graphically.

Although each of the roles in Figure 2.2 consists of just one 'thread' starting at the top of its grey box, a role might well consist of a number of separate threads corresponding to different things that it does. We shall see more of this later.

Representing new role instances being started – role instantiation

One role instance can *instantiate* another role, i.e. start a new instance of that role: this action is indicated by a square with a cross inside it. In Figure 2.2, the *Project Manager* role instantiates the *Designer* role. The caption against the crossed box identifies the role being instantiated (see Figure 2.3).

This idea of instantiating a role is of course a rather abstract one. In Chapter 1, I equated it with the idea of 'creating an area of responsibility', something separate from giving that responsibility to a real person, which, in Riva, we see as allocating an actor to a role instance – what we call 'casting', to follow the theatrical metaphor. We probably won't want to

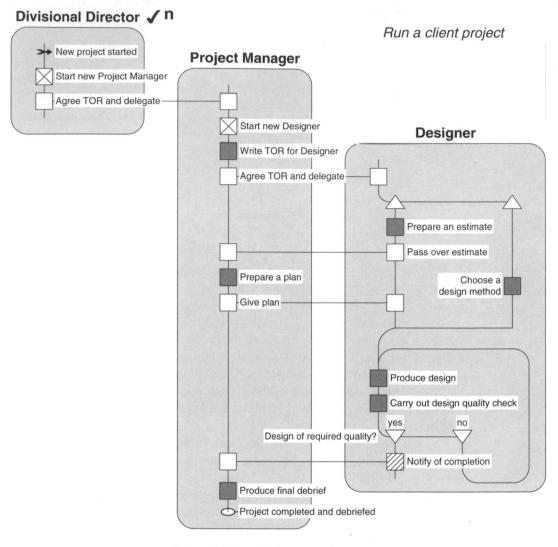

FIGURE 2.2 *A RAD for a simple process*

write *Instantiate Designer role* or *Instantiate the Task Force role* as the caption to a crossed box on a RAD. We are more likely to say things like:

- Start a Designer role.
- Create the responsibility for managing the project.
- Set up a Task Force.
- Get a Task Force going.

FIGURE 2.3 *Instantiating the Task Force role*

Strictly we should not say *Appoint a Task Force* as that really is about casting actors for the role instance. Indeed, whenever we instantiate a role, we have a modelling decision to make: do we want to model the casting of an actor to carry out the new role instance? Like all modelling decisions, the answer will depend on the purpose of the model. If we decide we do want to cover that aspect, we can expect the casting of the actor to be done after the instantiation: minimally an action such as *Nominate person to manage new project*, but potentially an interaction with other roles to come to a choice, or even with other processes responsible for resourcing.

In a RAD we have no separate symbol to represent the 'ending' of a role instance once its work is done. If we need to show this explicitly we simply use an action labelled something like *Close down this Task Force.*

Representing roles with pre-existing instances

We have seen that some role types have pre-existing instances: instances that are 'in place' when the process starts. For instance, we might expect that when the **Handle a Customer Complaint** process starts there is an instance of the role *Customer* already in place. It is important for us to distinguish those roles graphically for a very good reason: when we look at a RAD we want to be able to tell which roles have instances and hence where process activity can start. If a role does not have an instance in place when the process starts then there can be no activity to do with that role. The role must be instantiated first.

To mark a role with a single pre-existing instance we place a tick ✓ next to its name. If the role has exactly four pre-existing instances then we place the number *4* next to the tick: ✓4. If it has an indeterminate number of instances we mark it with '✓n'. See Figure 2.4. If a role has no tick against

FIGURE 2.4 *Roles with pre-existing instances*

its name we know immediately that, when the process starts, there are no instances of it and we can therefore expect to find it being instantiated by another role somewhere on the RAD.

Choosing and modelling roles

Different types of role

We know that roles come in different flavours. Does it make sense to have differently flavoured roles in the same model? The answer to this question, and to many similar ones, is 'Yes, if it makes sense'! When we model a process we have a purpose in mind. That purpose will tell us what makes sense. Let's start by looking at the different types of role and when we might use them:

- A unique functional position or post.
  Take the role *Chief Executive*. If we decide to have that role in our model we are clearly saying 'These are the responsibilities in this process that are placed on the desk of the person with the title "Chief Executive".' It's highly likely that the *Chief Executive* will have a host of other responsibilities in other processes. But a particular process model only shows their responsibilities in *this* process. The box marked *Chief Executive* in this model is not a full definition of the responsibilities of that post.

  Since such a role has only one instance we can expect that instance to be pre-existing and hence the role name will have an accompanying tick.

- A generic functional position or post.
  If all the Divisional Managers are in place when the process starts then we will put a tick against the role name. If we know there are six Divisional Managers then we can put a *6* next to the tick; if we don't care or only know that there is at least one Divisional Manager when the process runs then we put an *n* against the tick.

  If there are no Divisional Managers at the start of the process, then their creation must be part of the process.

- A unique functional group.
  The situation of a unique functional group is similar to the unique functional post: there will almost certainly be a single instance at the start of the process and hence a tick next to the role name.

- A generic functional group.
  Generic functional groups work like generic functional posts. An example might be *Retail Branch* in the process **Prepare the Annual Sales Forecast**. There will be a number – possibly indefinite – of them at the start of the process, and we shall tick the role name appropriately. This sort of functional group typically has a more-or-less permanent existence – unless of course we are looking at the process for deciding to open new retail branches and close existing ones.

- A generic type of person.
  Customer is the typical generic-type-of-person role. In most situations, the role will appear with a tick indicating a single pre-existing instance.

- An abstraction.

  Here the situation changes. Let's take an example: *Project Managing*. With abstract roles like this, we are naming the contents of the box: 'The set of responsibilities represented by this box we will call *Project Managing*.'

  The term 'Project Manager' is ambiguous in that it is used in different ways. Like some organizations, we might use it as a synonym for *Project Managing*: an abstract, transient role, whose instances only last as long as the projects they manage, each being associated one-to-one with the project. I could say 'I am the project manager of the Battlebridge Project.' On the other hand we might use it as a badge, identifying people qualified to act instances of the role *Project Managing*: I might say 'I'm a Project Manager' meaning that I get given projects to manage.

  A more pointed example might be *Large Claim Approving*, where we are naming the set of actions and interactions that carry out the responsibility of approving large claims. We have abstracted the role away from the organization and its structure – we are not saying what post a person must hold to do these actions and interactions, or what qualifications they must have – we are simply labelling the responsibility.

  An important abstraction that we shall see more of later is the sort that is also transient in the same way that a Project Team can be. Suppose I am a customer and make a complaint. In essence, I generate a responsibility to handle my complaint. It is very common in Riva to identify that responsibility as a role which is instantiated when a complaint arrives, deals with that and only that one complaint, and vanishes once that complaint has been dealt with. If we have 127 complaints being dealt with at a particular moment then we will find 127 instances of the role *Complaint Handling* in the organization. The moment a complaint is closed there will be 126 instances and if two new ones arrive, the count will go up to 128.

  Abstract roles might or might not have pre-existing instances on a RAD.

(Ticking pre-existing roles is obviously important when it is not obvious which roles have pre-existing instances and when we need to know. In the many snippets of RADs in this book, I have not always used ticks, in particular where the situation is obvious or does not matter.)

Abstract and concrete roles

The last type of role in the above list – the *abstract* role – is important. We saw in Chapter 1 that such an abstract role is almost a definition of the responsibility itself, rather than a label of a post that typically gets to carry out that responsibility. In a sense, we are getting closer to the role itself when we think of it in such abstract terms. A recurring theme will be the difference between our *intent* and the *mechanism* we use. We can model a process in terms of intent or of mechanism. For instance, if you ask me what I am doing I might answer 'I'm pressing keys on a keyboard': it's true, I am. But you might have expected the answer 'I'm writing a book.' I'm doing that too ... and the mechanism I am using is pressing keys on a keyboard. My intent is to write a book; I am doing it by pressing keys.

This distinction crops up when we choose the roles in a RAD. Do we want to talk about mechanism or intent? Is the mechanism that the *Finance Director* is the post that actually approves the payment of large invoices? Is the intent that the role executes the responsibility for *Approving large invoices*? It is probably, both so we have a modelling decision to make: *Finance Director* or *Approving large invoices*? The answer as ever will be obvious as soon as we remember why we are drawing this RAD. If we are preparing a process model as a work instruction then we had better be very specific about who does each job: we shall choose *Finance Director*. If we are trying to get inside a process and model what is really going on, or we are designing a new process and have no preconceptions about who does what, we shall choose *Approving large invoices*. But we shall see more of this decision and its answers in later chapters when we look at using Riva in different situations.

Committees and meetings as roles

It's often the case that a corporate body such as 'The Board' acts as a single role that performs various actions – monitoring, acting as an approval authority, planning etc – whilst at other points in the process the individual members act in their own right: CEO, CFO, CIO etc. The Divisional Directors might act collectively in one role – *Divisional Management Committee* – and individually in their own right in the role *Divisional Director*. In the latter case we might expect a Divisional Director to have an interaction with the Divisional Management Committee in order to submit a divisional plan as input to the corporate plan. If Shirley is a Divisional Director she will act both of the roles in that process, putting on the right hat at the right moment. She has two hats: one as *the* actor of *an* instance of the role *Divisional Director* and another as *an* actor of *the* single instance of the role *Divisional Management Committee*.

Equally, it is sometimes useful to regard a regular meeting as playing a role in a process, particularly if it has some executive responsibility. A meeting might be ordained to happen monthly in order to agree the priorities of the coming month's activity, to approve the expenditure of a department, or simply to ensure an exchange of information between the attendees. Such a role, e.g. *Monthly Planning Meeting*, might therefore appear on a RAD with a ✓.

People as roles

In the same way that a named department can be seen as a role, so can a named person. If we are modelling the **Respond to Customer Complaint** process in a very concrete way and Mary is the one who is responsible for calling customers who have left messages, then we can feel quite at ease equating her with that responsibility. On the other hand if we wanted to 'stand back' from the process and take a rather more abstract view of it, then we would probably not want to equate responsibilities with their current actors, and the role called *Mary* might appear as *Customer Recall*, say.

Computer systems as roles

We saw earlier that a computer can be an actor of a role (instance): in the work of the Accounts Department we will find people doing things (handling purchases, invoices, orders, cash advances etc), but we might also find computers doing things such as automatically preparing lists of aged debtors each Monday. It may be that an information system running on a computer plays such a large part in a process that we could consider it as having a role of its own. There would be one instance of that role and, trivially, one actor: the box of tricks itself. Other people-acted roles would of course have interactions with it, either to put data in or get data out. There is a sense of course in which we cannot really say that the computer system 'takes on the responsibility' of doing whatever it does, but showing a significant system as a role can be a useful modelling choice.

Although we have not yet fully explored the RAD notation, Figure 2.5 should be readily understandable and it shows a computer system, which is the role *Admissions Register System*, with Clinicians and Admissions Clerks interacting with it.

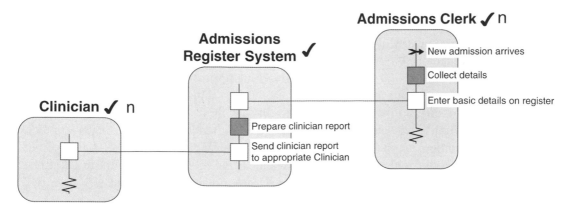

FIGURE 2.5 *A computer system as a role*

REPRESENTING ROLE STATES

The vertical lines between blobs *within* a role are more than just ways of connecting the blobs. They represent *states* which the role can be in. Our understanding of a RAD is greatly enhanced if we see lines as states rather than just 'flows' from one action or interaction to another. This will become more and more apparent as we look more carefully at RADs and the way processes work.

Sometimes on a RAD we want to say what state the world is in at a particular part of the process; in other words we want to label the state. We do this by simply putting a little 'sensing loop' around the state line and annotating it, as in Figure 2.6. Similarly, in Figure 2.2, the final state of the

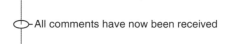

FIGURE 2.6 *Labelling a state*

Project Manager role is *Project completed and debriefed* (which is probably the goal of the whole process).

In real life we are quite used to the notion of state, even though we might not recognize it: 'How are you getting on with my expense claim?' is another way of saying 'What state has the processing of my expense claim reached?' We happily ask 'Has authorization to proceed been given yet?', or 'Has the Finance Director given his approval yet?' In these too, we are asking about the state the process has reached. And someone answering the question will say something like 'Well, it has reached the desk of the Finance Director but she hasn't had a chance to approve it yet,' or 'It's waiting for the Divisional Manager to provide some further figures.' We typically talk in terms of how far each role has got with the matter, especially if things have come to a halt on someone's desk!

Choosing and modelling states

History and potential

The first thing to say is that we do not need to label *all* the states in a RAD. This would be impractical and not useful. As ever, we label the ones we want to label, those that help us with our purpose in drawing the model.

We would label the *starting* state of a thread if there is something about the initial state of a thread that is important to the role or to its understanding: *Written complaint in hand*, *Nothing yet recorded on the database* or *The project has already been approved*. Some states in the middle of a thread in a role can also be important in some way: *All necessary materials now in hand*, *The file can now be closed* or *Everyone must now be informed*.

Note how some of these state descriptions say where we have been, and some say where we are going. In other words, we might summarize things up to this point: *All necessary materials now in hand*. Or we might say what is now possible or required: *The file can now be closed* or *Everyone must now be informed*.

These dual aspects of a state – history and potential – reflect the fact that one action's post-condition (the past) can be another's activating condition (the future). When we label a state we might wish to signal either history or potential, or both.

Goals as desired states

In Chapter 1 we saw how goals are important in modelling processes. A goal is simply a desired state. The goal is reached when the state is reached.

Let's take the example of a process called **Handle a Request for Quotation** in an insurance company. When a customer requests a quotation for insuring a particular risk, the (main) goal of the process is to provide that customer with that quotation. We shall have achieved that goal the moment the customer has the quotation in their hand. So in the role *Customer*, we should expect to find the state *Quotation in hand*, say, at some point.

We might be tempted to think that this marks the 'end' of the process: the main goal has been achieved. However, it might only be the end of the process as far as the customer is concerned. There may still be work to be done at the insurance company's end: database records to be updated, archives to be made, audit trails to be secured, information about the quotation to be added into the overall risk profile of the company, and so on. We might think of each of these as a 'minor' goal of the process: a further state that must be achieved before the process as a whole can stop. As such, it would be appropriate to label the states that correspond to these minor goals – they will mark the 'ends' of various threads of activity that are raised by the single request.

These major and minor goals are *point-wise goals*: we can identify the points (states) in the process where they are deemed to have been achieved.

As we saw in Chapter 1, there is a second sort of goal: the *steady-state goal.* Let's take as an example the process **Bring a New Product to Market** in a software product company. We could imagine that a goal of this process is that the Marketing Department always knows the latest key features of the product. This is a steady-state goal: we want it to be true all the time. By definition, therefore, we cannot put our finger on a single point – or state – in the process model where this goal is achieved, so it is meaningless to try to *model* such a goal. Instead, we will want to ensure that the process maintains that goal by *design*. This means knowing what in the process can perturb things (e.g. the software designers making a major change of functionality), how we could detect whether the Marketing Department was out of date, and how we could correct things if they got out of date. This is of course a design issue rather than a modelling issue, so we shall pick it up again in Chapter 11.

Multiple outcomes of a process

When we come to examine the logistics of process modelling in Chapter 8, we shall look more carefully at the importance of establishing process goals at the start of a modelling project. Indeed, we shall widen the question to one of establishing process *outcomes*, some of which might not be goals. For example, we might have a process to develop a new pharmaceutical drug and take it to market. The *goal* could be said to get the drug to market, but more often than not the actual outcome is that the drug is withdrawn and the project is closed. Project closure was not the goal of the process, though finding out *as quickly as possible* if a drug will not be successful can be, in one sense, a goal.

So when we look for outcomes we may well find several for a single process. Let's take the example of a process to **Handle a Customer Complaint** for a retail store. We might brainstorm several possible outcomes:

- The customer receives replacement goods and agrees the complaint is closed.
- The customer receives a refund and agrees the complaint is closed.
- No agreement can be reached and the matter is referred to the industry ombudsman.

So, we would expect to see each of these somewhere in the model of the process:

- The first outcome would correspond to the state after the receipt of notification from the customer that they have received the replacement goods and agree the complaint is closed. This state might perhaps be in a role such as *Customer Service Assistant.*

- The second outcome would correspond to the state after the receipt of notification from the customer that they have received the refund and agree that the complaint is closed. This state might perhaps be in the same *Customer Service Assistant* role.

- The third outcome would correspond to the state after the matter has been referred to the industry ombudsman. We might find this state in a role such as *Customer Service Manager,* say.

KEY POINTS

We mark interesting states with appropriate captions.
A state can express history (where we have been) or potential (where we can go).
Some states represent goals, sub-goals, or outcomes.

REPRESENTING ACTIONS

An action in a role is modelled with a small black box, suitably captioned, as in Figure 2.7. The action can start when its activating condition is met.

FIGURE 2.7 *An action in a role with its adjoining states*

When it is finished its post-condition is true. In Figure 2.7 we have shown the activating condition and the post-condition, by annotating the states before and after the action, but remember that we shall only caption states where it's useful.

The fact that an action is shown as a black box is significant: it says that, as far as we are concerned *in this model,* we do not care how this action is carried out, so long as the desired state is reached after it. The question naturally arises as to whether one can 'decompose' or 'open up' a black box. This is an important issue which we shall cover in Chapter 4.

'Running' a thread

When a role is instantiated we can think of the new instance starting with a 'token' sitting on the initial state of each thread. As the process unfolds and the role instance proceeds through its actions and interactions, the changes of state are marked by changes in the positions of tokens on the states, what we shall call the *marking* of the model (to borrow some terminology from Petri Nets). For instance, suppose a role instance is in the state shown in the left-hand fragment of Figure 2.8.

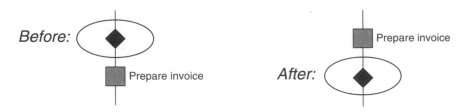

FIGURE 2.8 *A role instance thread before and after an action*

The token shown as a lozenge sitting on the state line before the action *Prepare invoice* indicates that the next thing that the role can do is carry out that action. Nothing else is needed for the actor to start the action. The state in front of the action represents what we have called its *activating condition*, that is, the condition of the role instance which will cause the action to start (strictly, to be instantiated). A token sitting on a particular state can be thought of as representing the potential future behaviour of the role instance.

When the action does start, the role instance is in the state of carrying it out (and the token essentially disappears, though you might think of a token sitting in the action box). When the action has finished, a token appears on the state immediately following the action to indicate that the role instance is now in that state. The state after the action represents what we have called its *post-condition.*

We shall use this idea of tokens flowing to illustrate what happens when a RAD 'runs'. It is important to understand just what happens when a RAD 'runs' so that we can make correct interpretations of the model for comparison with what happens when the real-life process runs, or to see how a proposed process would look when it does run. Above all, it will help us check the concurrency that we have modelled, both within one role (instance) and across the process as a whole. This is one of the features of RADs that distinguishes them from flowcharts and swim-lanes, where serial threads are the order of the day. We shall examine process concurrency in detail in Chapter 3.

Choosing and modelling actions

Actions as black boxes

An action is shown as a black box and it is useful to think of it in exactly those terms: 'This action is atomic; we are not saying how it is done or what is in the box.' Anything we do wish to say about the action appears in the caption we attach to it. We are not likely to write an essay as a caption but we might have a sentence or two. For example we might write:

- Get deployment document, operational review, and support documentation signed off.
- Update file.
- Prepare annual forecast.
- Decide whom to involve in forthcoming project review.

When we choose to 'bottle up' some activity and represent it as a single box we are making an important modelling decision about the amount of detail that we want to get into. *Get deployment document, operational review, and support documentation signed off* may be regarded as atomic for a particular RAD but it is of course potentially a 'big' complex action. My diatribe against hierarchies in Chapter 1 will have warned you that I am against decomposition as a basis for modelling. So, I would be very, very cautious about wanting to 'decompose' a black-box action – exactly what does 'decomposing an action' mean? This is in fact a very difficult question, one that process modelling methods that use decomposition ignore, with the result that they falsify their models. The question is important enough that we shall leave it to its own treatment in Chapter 4. Meanwhile, in answer to the question 'How much detail should I go to in my process model?' I would simply answer – of course – 'However much is useful given the purpose of your model.'

Concrete and abstract actions

I pointed out the difference between intent and mechanism earlier on, when discussing roles. A similar view can be taken of actions: we can model either the intent of the action, or the mechanism used to carry it out, or both. For example, we might caption an action as *Complete screen 24A* – this gives us no idea at all of what is intended by this action but it does tell us what to do, what mechanism we should use. On the other hand, we might caption the same action *Record new customer details*, which tells us what we are trying to achieve with this action – a record of customer details – but gives us no indication at all of how to do it. The first caption treats the action in a concrete sense, the second in an abstract sense (we are abstracting away from mechanisms to what is intended). Of course, we might decide that the purpose of our model is best served by putting both the intent and the mechanism in the caption: *Record new customer details using screen 24A*. A purely mechanistic caption might be sufficient in a RAD serving as a work instruction or procedure. A purely abstract caption might make sense if we are using our model to really get an understanding of what the process is all about, irrespective of how we go about it physically. The combined caption could be used in training material where it is good to tell people what to do and why they are doing it. As ever, a process model must be fit for purpose.

Actions need verbs

Finally, note how the caption for an action starts with a verb. It is activity that we are describing after all, so it's a good discipline to ensure that there is an active verb somewhere.

> **KEY POINTS**
> **An action is shown as a black box.**
> **An action is atomic for the RAD.**
> **An action has an activating condition and a post-condition, and we might choose to model them.**
> **An action can be described in terms of its intent or its mechanism, or both, as appropriate.**

For brevity, I shall sometimes refer to the *pre-state* and the *post-state* of a process element, meaning the activating condition and the post-condition, respectively.

REPRESENTING CONCURRENT THREADS OF ACTIVITY

There might be a point at which a role can start a number of separate threads of activity that can be carried out concurrently. 'Now that I have written all the chapters I can prepare the contents list and at the same time I can prepare the index.' This 'splitting' of a thread into two or more concurrent threads is represented in a RAD by the symbol shown in Figure 2.1 for *part refinement*. Strictly speaking, a state of the role is being refined (divided) into a number of separate parts. Another example of part refinement is the early split in the *Designer* role in Figure 2.2 where a designer starts two concurrent threads of activity. On one thread they choose a method; on the other they first prepare an estimate, then interact with the project manager to pass over the estimate, and finally wait for a second interaction to receive a plan back from the project manager. Part refinement can involve any number of threads of concurrent activity, depending on just how much concurrency is possible in the work of the role.

Using tokens again, we can think of the single token that reaches the part refinement becoming a number of tokens, each of which passes down one thread of the part refinement. In the left-hand fragment in Figure 2.9, the action *Do Z* has completed and there is a state token on the state line coming out of *Do Z* and before the part refinement. This marking is entirely equivalent to that shown in the right-hand fragment, where the token before the part refinement has 'turned into' one on each of the separate part threads.

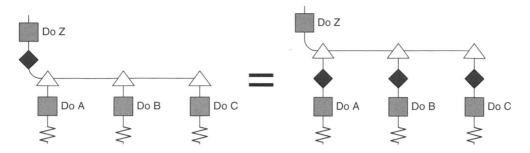

FIGURE 2.9 *The marking before a three-way part refinement*

In some situations, all the concurrent threads must complete before the role can proceed to further activity; in this case we use the representation in Figure 2.10, with the four threads being joined once they have finished (i.e. the part states are recombined).

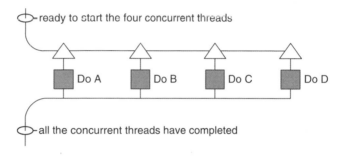

FIGURE 2.10 *Closing a four-way part refinement*

We can expect that at some point after the part threads have gone their separate ways, they will all have finished and hence there will be a token sitting on the state at the end of each thread, as shown in the example in the left-hand fragment of Figure 2.11. This is precisely the same marking as

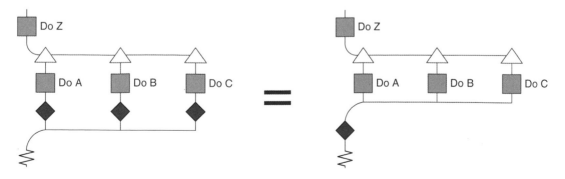

FIGURE 2.11 *State recombination at the end of a three-way part refinement*

that shown in the right-hand fragment, where the part thread tokens have all been replaced by a single token on the state line immediately after the closure of the part refinement.

In some cases, however, a role does not operate this way. A project manager's activity might be considered as two quite separate areas of responsibility, almost 'sub-roles': keeping staff occupied with work and liaising with the client. These might be the two threads of an early part refinement of the role, which need never recombine. In Figure 2.12 we see

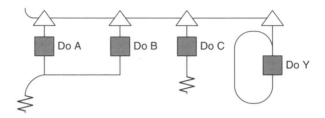

FIGURE 2.12 *A four-way part refinement where only two threads recombine*

a fragment in which two of the four threads recombine, whilst two others never do, one endlessly looping on itself and the other diving off to somewhere else in the role perhaps.

Replicated part refinements

Suppose a Line Manager wants to prepare a revenue prediction across all the projects they are responsible for. The number of projects they have active at any one moment may be variable, so we need a way of representing this. This is one example of a general situation where we want a thread of activity to be 'replicated' a number of times. Figure 2.13 shows how we do it.

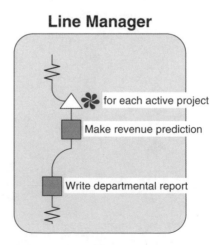

FIGURE 2.13 *A replicated part refinement*

The single thread that is to be replicated is shown within the usual part refinement structure and the replication is indicated with an asterisk. The number of times the thread is replicated is captured in the caption, in this case *for each active project*. Because this is a part refinement, the action *Write departmental report* cannot start until all the replicated threads are complete, i.e. until the Line Manager has made a revenue prediction for all the current projects.

Note that *this is not a loop*: the Line Manager is *not* making a revenue prediction for one project, and then for the next, and then for the next, until they are all done, in a serial fashion. Instead we are saying that there is a thread of activity – a simple one in this case consisting of just one action – which is replicated, say seven times, and all those seven threads can now start and run in parallel. Figure 2.14 shows the replicated part

Line Manager

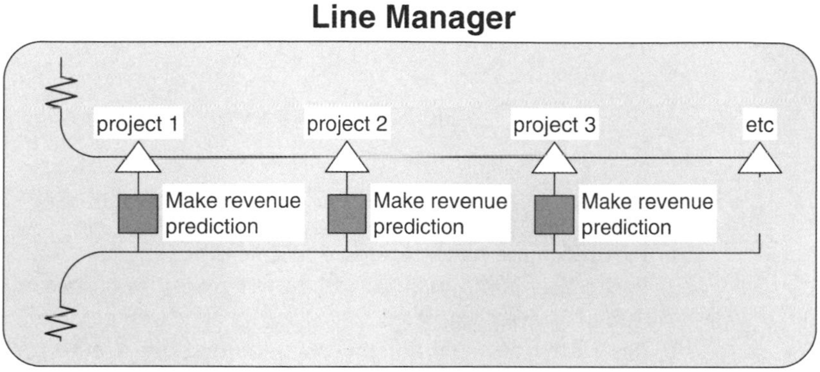

FIGURE 2.14 *The replicated part refinement in Figure 2.13 expanded*

refinement expanded. As in a normal part refinement, the replicated threads might or might not recombine.

It is important to remember that the replication is done the moment it is reached. On one occasion a thread might be replicated seven times and on another, 70 times.

Choosing and modelling part refinements

The part refinement is an important construct because it is one way we model potentially concurrent action within a role. When people model processes, they all too easily fall into the trap of making all activity *sequential*: A follows B follows C follows D etc, thereby ending up with a very sequential-looking process. If we are modelling an existing process, the resulting model might be true in that what we have drawn does capture the *time ordering* of actions, but it might not model the *necessary logic* of the process. Suppose action *B* must follow action *A*, and that action *D* must follow action *C*, but that those two threads – *A–B* and *C–D* – are independent. Because I can only do one thing at a time, you might observe me doing any one of the following sequences: (*A*, *B*, *C*, *D*), (*A*, *C*, *D*, *B*), (*A*, *C*, *B*, *D*), (*C*, *D*, *A*, *B*), (*C*, *A*, *B*, *D*), or (*C*, *A*, *D*, *B*). But none of those is the only or correct way. It would be wrong for us to observe things being done in the order *A*, *C*, *B*, *D* one day and then to show that sequence on our model. It would be an incorrect model of the process in general. And if we were *designing* the process and we drew it as in Figure 2.15 we would unnecessarily restrict people's options for carrying out that process in the future.

FIGURE 2.15 *An over-constrained process*

KEY POINTS

Part refinements capture within-role concurrency.
Part refinements might or might not recombine once some or all of the threads have completed.
Replicated part refinements model situations where the same thread is carried out a number of times in parallel.

REPRESENTING ALTERNATIVE COURSES OF ACTION

At some points in a process, what happens next in a role might depend on the state reached. For example, the way a clerk deals with an application for overtime might depend on the salary band of the claimant; how a chemist makes a batch of drug compound for a clinical trial might depend on which pilot plant has been allocated for the production; the way an order for a shrub is dealt with by a horticulturist might depend on the time of year the order is received and when the shrub concerned is best shipped.

We represent such alternative courses of action with the notation shown in Figure 2.1 for *case refinement*. Essentially we are refining the state of the process according to different 'cases'. The general situation is shown in Figure 2.16.

The preceding state line takes a bend to the right and a downward pointing arrow appears at the start of each alternative thread. Typically, we label the bend with a question:

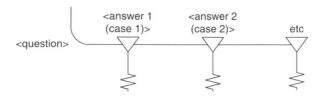

FIGURE 2.16 *Representing alternative threads of activity: case refinement*

- Has the invoice been paid?
- What month is it?
- Where is the package going?

We then label each downward arrow with an alternative:

- yes/no;
- January/February/... ;
- UK/overseas.

Using our token scheme, we can think of a token arriving at the case refinement, and then passing down the thread that corresponds to the predicate that is true; the role goes in different directions depending on the state of things at that moment. Figure 2.17 illustrates this. Immediately

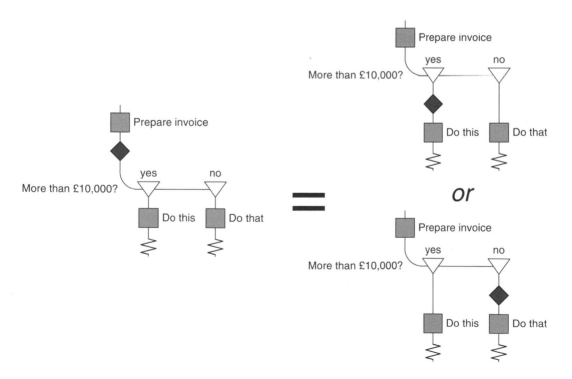

FIGURE 2.17 *The marking before a case refinement*

after the action *Prepare invoice* we can imagine a token on its post-state as shown on the left-hand fragment. The case refinement says that if the predicate *More than £10,000* is true, this is equivalent to the upper right-hand fragment with the token on the corresponding state line; whilst, if *More than £10,000* is not true, it is equivalent to the lower right-hand fragment with the token on the other state line.

The two-way case refinement generalizes quite naturally to *N*-way case refinements. The way an organization carries out a particular part of the process might, for instance, depend on which of its offices it is carried out in. We would show this with an *N*-way case refinement such as that in Figure 2.18. Here, the process proceeds differently according to whether the location is London, New York, Paris, or Stykkisholmur.

FIGURE 2.18 *A four-way case refinement*

You could think of case refinement as a *case* statement in a programming language or as a decision box in a conventional flowchart. But it is important to note that, unlike a decision box on a flowchart, there is no activity going on 'in' the symbol for case refinement – no person or machine is doing anything to *make* the decision: the role instance is simply going in different directions depending on the state it is in. The upper fragment in Figure 2.19 would therefore be wrong: the caption is a

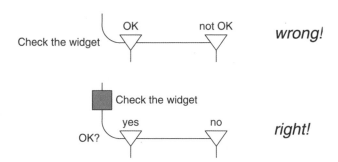

FIGURE 2.19 *Nothing 'happens' in a case refinement*

description of an action and not a predicate (question) about the state. The value of the *OK?* predicate must be determinable as a result of some prior action or interaction in the process, such as the preceding quality control action *Check the widget* in the lower fragment; the case refinement does not itself 'contain' any activity to check the design.

In some situations, whichever of the alternative threads of activity is followed, we want to 'return' finally to the 'main' thread of activity. In this case we use the representation in Figure 2.20, with the threads joining up

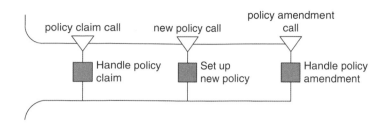

FIGURE 2.20 *Case refinement threads sometimes recombine*

again when they have finished (i.e. the case states are recombined). In this example, we have also used an alternative labelling of the case refinement: we have omitted the question and simply labelled the different cases; if the meaning is clear, this is fine.

We can best visualize what this means by looking at it using tokens. In Figure 2.21, whichever of the three case threads is followed, we require the

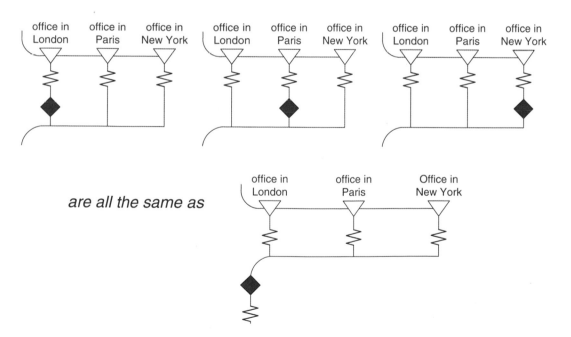

FIGURE 2.21 *Three case refinement threads finally return to a common thread*

main thread to be picked up finally. Thus each of the three markings shown in the upper part of the figure is equivalent to the marking in the lower part.

In other situations, the case refinement threads might not recombine. The example in Figure 2.22 is a case in point: if the customer call picked up

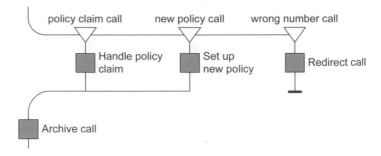

FIGURE 2.22 *Case refinements don't always close tidily*

at the call centre is about making a claim on an insurance policy or setting up a new policy, then it is dealt with, and once it has been dealt with it is archived. If the call is a wrong number, we simply redirect it and finish the process there. (If this all seems hopelessly laboured and obvious, then you are probably not a software engineer: in software design and modelling methods that have come from that world, there is a strong emphasis on 'Dijkstra structures' where, in crude terms, everything closes off tidily. Unfortunately the world is not as tidy and 'block-structured' as we can make our software, so any modelling method that demands such tidiness will be of no use in modelling the real world.)

We sometimes adopt an abbreviation for simple case refinements. On the left-hand side of Figure 2.23 we show a case refinement with a single

FIGURE 2.23 *Abbreviating a simple case refinement*

action only being done if cleaning is necessary, whilst on the right-hand side we capture the conditionality in the name of the action. Words like 'if appropriate' or 'as necessary' might be useful. This can also be done with conditional interactions, but greater care is needed.

Case refinement naturally allows us to represent conditional iteration within a role, as in the example in Figure 2.2 where the Designer repeatedly produces and checks a design until it is OK. Anywhere on a single state line is the same, so the three role instance markings shown in Figure 2.24 are equivalent in that the next possible action of the role is *Produce design* in

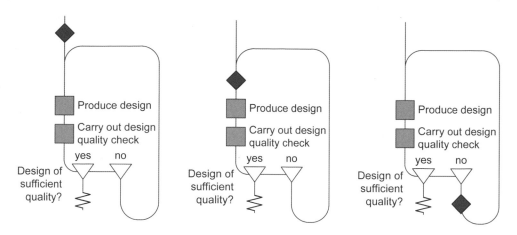

FIGURE 2.24 *Three equivalent role instance markings*

all three cases. You might object that the first and third situations in the figure are quite different, since in the first you don't have a design whilst in the third you do. True, but the RAD is telling us that, as far as this process is concerned, having a design that has failed its quality check is no different from having no design at all: in each case you have to produce a new design. The states marked by the tokens are equivalent in that *they define the same future behaviour* even if they define different histories. If this is not what we wanted to model then we have the wrong model.

Choosing and modelling case refinements

Case refinements are not active

Case refinements are relatively straightforward. But we should always make sure that all the information needed to evaluate the predicate is available to the role concerned without further work. The more general question we should ask is 'Does the role have all the props necessary to answer the question?'

So if the predicate is *Which office is the application being processed in?* there probably isn't a problem: the actor can look around and recognize they are in New York and not London. If it is *Is the applicant male or female?* and the applicant isn't sitting in front of the role concerned, then we should check whether the applicant's gender has been ascertained before this point in the process, and that it is available to the role concerned.

Complex case refinements

There are situations where we can model the case refinement in several ways. Take the following description of what happens in the pharmacy of a pharmaceuticals company:

A pharmacy bottles three types of drug:

1. unformulated and aseptic;
2. unformulated and non-aseptic;
3. formulated.

Type 1 requires equipment to be sterilized before filling. Type 2 requires equipment to be cleaned before filling. Types 1 and 2 require bottles to be tested for leaks after filling. Type 3 only requires bottles to be filled.

We have several options, shown in Figure 2.25. They all describe exactly the same behaviour: for each type of drug the same actions are done in the

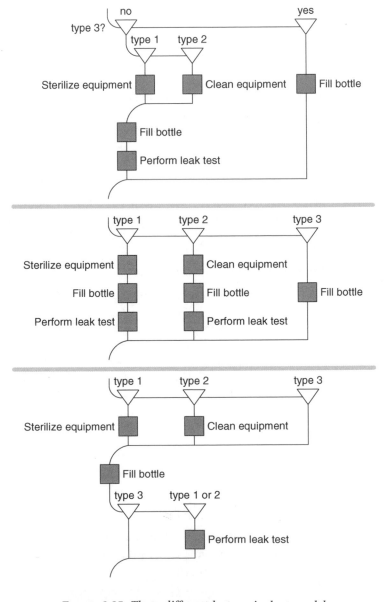

FIGURE 2.25 *Three different but equivalent models*

same order. So, is there any reason to use one rather than the others? The first option might be appropriate if type 1 and type 2 drugs were prepared in one location and type 3 in a different location; the way that the threads are placed suggests the 'geography' of the process. The second option might be appropriate if the three types were prepared in different locations, so it is useful to elaborate the thread fully for each type. The third option might be appropriate if all three types were done in the same location and one wanted to emphasize that bottle filling was done in the same way (one single action) in all three cases.

Probabilities and case refinements

Case refinements represent alternative courses of action, and when we come to analyse the process it might be important to know how often the different alternatives are followed. In Figure 2.26, how often does the

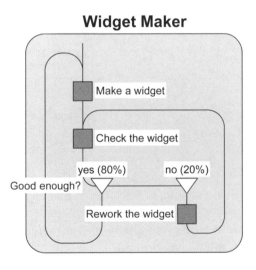

FIGURE 2.26 *An annotated case refinement*

Widget Maker have to rework the design? We simply annotate the RAD in an appropriate way. Figure 2.26 shows a simple percentage, but there might be seasonal variation or it might depend on some other characteristic of the process (the depth of the items in the in-tray, the rate at which work is being done etc).

> **KEY POINTS**
> Case refinements model alternative courses of action that arise in a thread.
> The alternatives should be mutually exclusive.
> Each alternative leads to a separate thread.
> The alternative threads might or might not recombine.
> There is no activity in the case refinement itself.

Representing merging threads

It is quite often the case that we want two or more threads in a role to come together in a single thread, even though they did not originate from a single thread. For instance, the Payroll Department will prepare a cheque for an employee and get it authorized at the end of a number of different procedures, all of which start from different triggering points:

- At the end of the month when the salary payment is due and the timesheet has been checked.
- Whenever a cash advance has been approved.
- Whenever the reimbursement of expenses has been approved.

To show these different parts of the role coming together to a single thread we simply combine the state lines in the curving fashion shown as in Figure 2.27. This says simply that whichever of those states the role is in, its

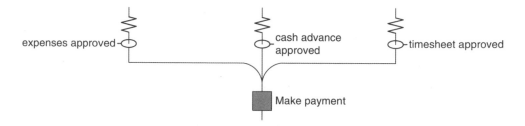

FIGURE 2.27 *Three threads combine to form one*

future behaviour is the same: it makes payment. Each of the three threads defines a different history, but they all also define the same potential. (Note the different shape in the merge from the way that part and case refinements are closed.)

The threads can come from anywhere within the role. Organizational activity often ambles around, dividing and recombining, ducking and weaving, jumping off at tangents, sometimes coming back. Our notation must allow us to model the 'untidiness' of the real world. Within a role, threads of activity can divide, recombine, and switch to other threads without constraint – simply because that is the way the real world operates: as a network rather than a hierarchy. And roles can operate in a similar way using interactions as the mechanism for 'jumping'.

Representing the end of a thread

Do we need to signal the end of a thread? No. The left-hand fragment of Figure 2.28 shows a thread simply coming to an end. Once the action *Archive paperwork* has been completed there is nothing else to do and so nothing more will happen on this thread. In some situations however, it can be helpful to reinforce the fact that this is really the end of the line and

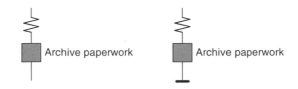

FIGURE 2.28 *The optional stop sign at the end of a thread*

not simply as far as we wanted to go in this model. In these situations we use a little 'stop' sign as in the right-hand fragment.

KEY POINTS

States in a role can merge: different past behaviours lead to the same future behaviour.
Optionally, we can mark the end of a thread.

REPRESENTING INTERACTIONS

Interactions are collaborative actions carried out by two or more roles. Let's take some examples:

- The purchaser arranges finance with their bank.
- The Project Manager and Line Manager review the Project Plan.
- The QA Team lets the Project Manager know they have finished the tests.
- The Line Manager asks for an update on progress.
- Vendor and Purchaser agree on the delivery date.
- The Sales Executive tells the Project Team of the contract change.
- The Product Design Team gives the Development Team the specification.
- The Engineering Department drafts the design with the Production Engineer.
- All the Divisional Managers meet to review the budgets.
- Each Divisional Manager reports to the Board.

A vanilla *interaction* between roles is shown as a white box in one role connected by a horizontal line to a white box in another role. We refer to the white box in each role as a *part-interaction* – it is the contribution that the role concerned makes to the interaction. Figure 2.29 shows a simple interaction between two roles.

In this example, by shading in its part-interaction, we have also shown which of the roles 'takes the driving seat' and makes the interaction happen. We are saying that it is up to the widget-making supervisor to take

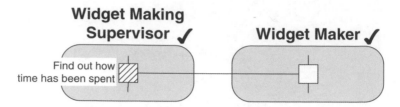

FIGURE 2.29 *A simple interaction*

the initiative and ask the (single) widget maker how they have been spending their time.

An interaction can involve any number of roles and signifies that the roles involved must pass through it together; as an example Figure 2.30

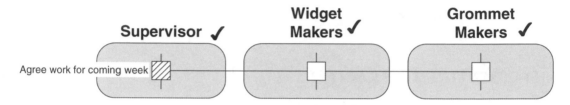

FIGURE 2.30 *A three-party interaction*

shows an interaction involving three roles. Note also how in this example we have lumped all the widget makers as one role, *Widget Makers*, and all the grommet makers as another, *Grommet Makers*.

If it is useful, we can add further annotation to the interaction, perhaps captioning a part-interaction to describe its contribution and even a further caption for the entire interaction. An example is given in Figure 2.31. We would typically model in this way when capturing a truly

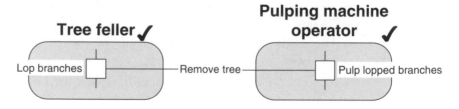

FIGURE 2.31 *Captioning the components of an interaction*

collaborative activity where each of the parties makes their own contribution to the overall action.

We always caption an interaction in a way that makes clear what is happening, and that would include whether anything 'changes hands' during the interaction, i.e. whether any grams are involved. Interaction lines do not carry arrows to indicate the 'flow' of any gram: we simply place an appropriate caption at the appropriate end. For example, in Figure 2.2 the *Project Manager* role receives an estimate from the *Designer* role, and this is indicated by placing the caption *Pass over estimate* at the *Designer* role end of the interaction line.

Interactions synchronize role instances

As with an action, an interaction has an *activating condition*, which is the condition corresponding to each participating role being in the state before its part-interaction. So the overall activating condition is effectively the state when all the participating roles are ready for the interaction.

The rules for an interaction say that an interaction cannot start until all the participating roles are in their respective pre-states, and that when the interaction finishes they all move into their respective post-states. Any role might get to the interaction before the others. If I reach my side of the interaction and you are not ready, I must wait until you are; as soon as we are both ready the interaction can take place. Put another way, *interactions synchronize the states of the participating role instances*. Figure 2.32

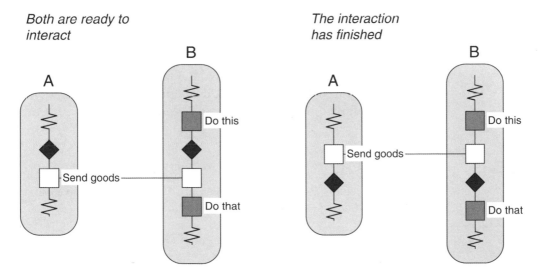

FIGURE 2.32 *Interactions synchronize role instances*

summarizes this with tokens. In the left-hand fragment both roles are in their respective pre-states, i.e. their respective part-interactions can be activated. In the right-hand fragment, the two part-interactions have completed and so the two roles are in their respective post-states.

How much do we want to synchronize?

Remember that a RAD shows strict ordering. It is very easy to draw strict ordering when it is not actually required or present in reality. For instance, if role *B* must have got to a certain point but then can accept the goods any time after then, *and* must have them before certain other actions can continue, then we render this as shown in Figure 2.33. This figure says that once B has

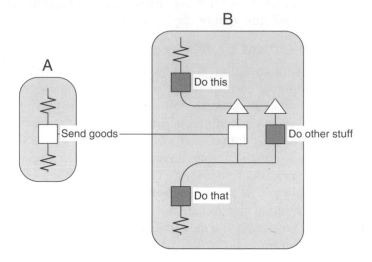

FIGURE 2.33 *Just enough synchronization*

completed the action *Do this* it can receive the goods while getting on with other tasks (*Do other stuff*) as shown by the part refinement. Once it has received the goods *and* completed those other tasks it can *Do that*.

We have recognized that there is a window during which B can and must take delivery but that during that window, B can be getting on with other things.

Later, next to the water cooler

Pupil:	You've emphasized that interactions synchronize state and ... if I've got this right ... an interaction therefore can't finish until all the parties have done their part-interactions?
Tutor:	Correct.
Pupil:	But surely that's not like the real world. Suppose I send you a letter. Sending you a letter is an interaction between you and me, but I don't wait around until I've heard you've received it before getting on with something else.
Tutor:	You have a knack of putting the answers to your questions in your questions! You're right that the interaction of my-sending-and-your-receiving-a-letter doesn't finish until – in particular – you've received it. But you told me that you don't want to wait around for the interaction to complete. In effect, you told me that you want to do something else concurrently while that interaction is completing. Doesn't that sound like a part refinement?
Pupil:	I guess it does. Shall I draw it on this handy whiteboard next to the water cooler? (Figure 2.34) I suppose that if I wanted to capture what we do once you have indeed received the letter then I would draw it after the interaction, but I've drawn

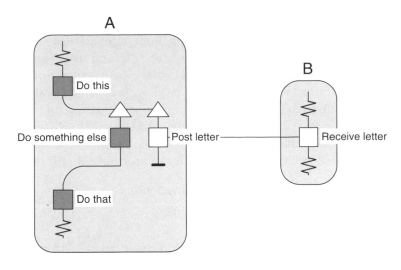

FIGURE 2.34 *Just enough synchronization again*

it as if once I've posted it, that's it. Of course I can see now that it's very like Figure 2.33 but reflected.

Tutor: Yes. Now, if we don't mind modelling mechanisms, we could recognize that when you post a letter you have a very short interaction with the post office immediately after which you can get on with other things. The post office then carries out the action of moving your letter from one post office to another, and then has a very short interaction with your correspondent to give them the letter. We have decoupled you and your correspondent through what I would call a carrier function. But you can draw that for yourself – I'm going home.

Choosing and modelling interactions

Interactions change role states

Interactions can come in a great variety of guises:

- I arrange finance with my bank.
- We review the Project Plan.
- He lets her know he has finished the tests.
- She asks for an update on progress.
- They agree on the delivery date.
- She tells the team of the contract change.
- I give you the specification.
- We draft the design.
- They report to the Board.

Reading these, we sense different things going on. Sometimes things are being passed over: information, documents, materials. Something that was in one role is now in another; or perhaps they both have it, as in the case of information. Sometimes, the interacting parties are jointly contributing to something: reviewing, agreeing, drafting. One or both of them now has something that neither had before the interaction: an agreed delivery date, for instance, or a design that they have jointly drafted. It is useful to think through the state changes in the participating roles: just what has changed in each role as a result of the interaction? How have the states of the props been changed? What props does the role now have that it did not have before? What props does it not now have that it did have before?

Concrete and abstract interactions

Interactions are as amenable to concrete and abstract descriptions as roles and actions are.

A caption on an interaction that read *Send completed form 195/5* would tell us nothing about what is going on except how the interaction is done. If we knew the form concerned we might know that it is the company's expense claim form and hence the interaction is in fact about claiming expenses. An abstract caption for the same interaction might be *Claim expenses*, which would not give us any indication of how to do it but would tell us what was going on, the intent of the interaction. The caption *Claim expenses using a completed form 195/5* would tell us what was going on and how it was done.

Here are some other examples:

- A Project Manager passes a budget report to the Line Manager ... A Project Manager reports on the budget to the Line Manager.

- The Client and Consultant speak on the telephone ... The Client and Consultant agree the scope of the work.

- The Customer presses the 'Buy Now' button on the web page ... The Customer confirms the purchase of the contents of their shopping cart.

When would we use these different styles of caption? Again, this is a discussion that we must leave until later chapters where we look at how Riva is used in different situations.

Interactions with replicated role instances

When a process runs, an interaction actually occurs between *instances* of the roles concerned, of course. (Strictly, an *instance* of the interaction occurs between instances of the roles concerned!) So, one instance of *Line Manager* interacts with one instance of *Project Manager*. But what if the Line Manager wants to have the same interaction with each of the current Project Managers, perhaps to get the status of their projects? We need to be able to represent the fact that the one instance of *Line Manager* has the same interaction with all the existing instances of *Project Manager*, however many there are. The upper fragment in Figure 2.35 shows how we

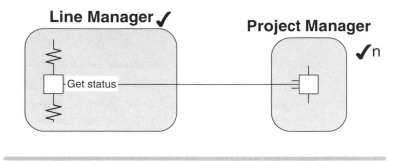

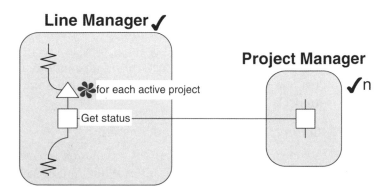

FIGURE 2.35 *A replicated interaction*

do this. You can think of this as shorthand for the lower fragment in Figure 2.35 in which the replication is shown at the *Line Manager*, end by replicating the thread containing the interaction for each Project Manager. The net effect is that the one instance of *Line Manager* starts a separate interaction with each instance of *Project Manager*.

Note that this is not the same as having a *single* interaction that involves all of them. An example might be that the Line Manager wants to brief all the current Project Managers on the new corporate budget. We show this using the notation in Figure 2.36. This single interaction involves a total of

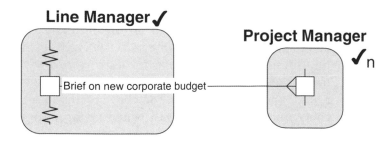

FIGURE 2.36 *An interaction involving many role instances*

$N+1$ role instances, where N is the number of instances of *Project Manager*.

Interactions are rarely as simple as we think

An interaction can be very simple (e.g. *I give you some terms of reference*) or very complex (e.g. *The three parties meet to negotiate and agree the price of a piece of work, drawing up the agreement as a legal document and obtaining financial securities from a bank*). As with actions, what we regard as 'atomic' in our RAD depends on why we are drawing the RAD. We always show whatever detail is appropriate to *that* model for *that* purpose. An interaction that is shown as a single line on a RAD might, if it were in some sense 'opened up', show the involvement of other roles not otherwise mentioned in this RAD, new interactions between them and other roles, and a whole mass of organizational activity. But by drawing just one interaction we are saying 'One interaction is appropriate for this model.'

Contracts – a pattern of interaction

A contract is a common form of interaction between two parties and one that we shall see in various guises. Winograd offers a view of organizational behaviour in terms of a four-step contractual cycle between a customer and a supplier: preparation, negotiation, performance, assessment.

In the preparation step, either the customer decides what they want to buy (or, more generally, contract out for) or the supplier makes some offer to a would-be customer; this results in the customer making a request to the supplier. The two parties negotiate the request and two mutual promises result: the supplier agrees to provide something in return for something else, perhaps payment by the customer. The supplier then carries out the performance step which finishes with a declaration that the work is complete, an assertion that the customer tests, finally declaring satisfaction.

Each step can, in its turn, be carried out by a cycle of its own: performance might be broken down into sub-cycles for instance.

In the ideal world conceived by Winograd and Flores, or a new one that we intend to build, we build all organizational behaviour out of such contractual cycles (see Winograd, 1987). It is doubtful that we can retrospectively *impose* a hierarchical structure of such cycles on an existing process or expect to find one there when we come to model it – we cannot expect that the evolution of the process has preserved such neat conceptual integrity; in fact, it would be far more realistic to expect that as in any other physical system, entropy – the degree of chaos or lack of order in the system – inexorably increases over time unless energy is expended to reverse it (albeit temporarily). We could of course consider BPR to be an 'entropy-reversal' exercise, one in which the chaos and disorder that has built up over the years is recognized and, through radical change, replaced by a 'tidier' and 'simpler' system.

Nevertheless, the Winograd contractual pattern can be observed, and of course has a natural representation in a RAD: Figure 2.37.

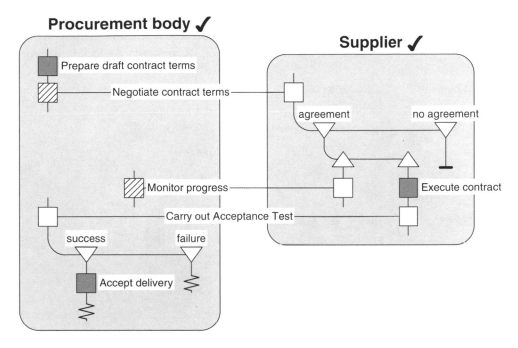

FIGURE 2.37 *A contractual cycle in a RAD, driven by the customer*

We can elaborate the model at several points to allow for situations where

- there is a breakdown in the negotiation;
- the *Supplier* fails to complete the performance;
- the *Procurement body* is not satisfied with the performance of the *Supplier*;
- the various steps are themselves the subject of subcontracts.

We shall find it a useful modelling discipline to spot interactions that represent acts of negotiation, assessment and so on, in order to see how far the complete cycle is present and perhaps, if it is not, to ask whether it should be and whether the process could be improved by restructuring it to the 'standard' cycle.

So, whenever, in a modelling workshop, someone identifies an interaction by saying 'Then the employee gets the Line Manager to approve the expense claim', we shall now know to ask 'Do they always approve it?' If it really were the case that the Line Manager always approves the expense claim, we could identify an easy process improvement by simply removing that interaction – it seems to serve no purpose. What is more likely of course is that a more complex conversation is being (poorly) summarized as one of approval, and we are probably ignoring the possibility that the Line Manager might reject the expense claim. And what happens then?

We can elaborate this idea a great deal further and recognize interactions as 'conversations for action', with the much more complex general pattern shown in Figure 2.38. The conversation between two

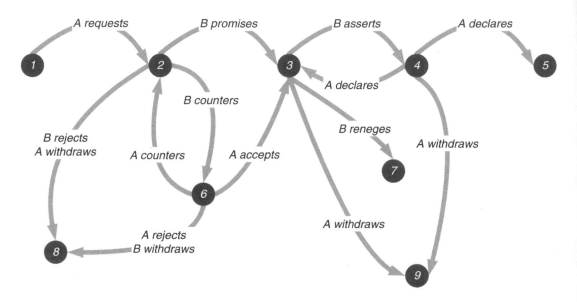

FIGURE 2.38 *An interaction as a conversation for action*

parties *A* and *B* starts at state 1 and then moves to a new state depending on how *A* and *B* interact. The interaction therefore ends up at one of states 5, 7, 8 and 9; the last three of these represent a form of failure, the first success. (See Winograd and Flores, 1987.)

This template gives us a useful pattern against which to assess each interaction when we come to it. I am not suggesting that every interaction we deal with should be elaborated to this *n*th degree, only that we should make a conscious decision about how far we want to go in *this* model, for *this* purpose, in unpicking *this* interaction into its component parts. Returning to our expense claim example, it would be entirely valid for us to wrap up a whole mass of such conversational detail in a single interaction that is indeed labelled *Get the Line Manager to approve expense claim*, if we allowed that that initial refusal, reworking, and resubmission were part and parcel of the interaction. That would still not allow the possibility of my withdrawing my expense claim altogether of course, and if that possibility was important and relevant to the model we were drawing, then we had better draw it and not simply ignore it.

Delegation ladders

When we model a process with a RAD, we seem not to take any explicit notice of one of the most important aspects of an organization: its authorization hierarchy. Most organizations – even those operating forms of matrix management – use some layering down from the Chief Executive; some *only* operate that way. In fact in a RAD, although we might not model the hierarchy explicitly, we can model the way it makes itself felt: the business rules that operate in terms of planning, delegating, reporting, authorizing, and so on. Indeed, it is normal for a RAD that uses functional positions or job titles as roles to expose the hierarchical aspects of an organization's behaviour in terms of interactions between 'superior' and 'inferior' roles. It is common to see the hierarchy running from left to right in the RAD: *The Board* appears as a role at the extreme left, passing instructions to the next level down, say *Divisional Director*, who in turn passes instructions to their right via interactions with *Project Manager* and so on further to the right. We can see these as formal contracts of course, as discussed above.

Delegation and reporting back are very common process patterns in an organization and they have a very natural representation in a RAD: they are simply pairs of interactions. Take the process shown partially in Figure 2.39. If *The Board* starts up a *Customer Survey Task Force* to carry

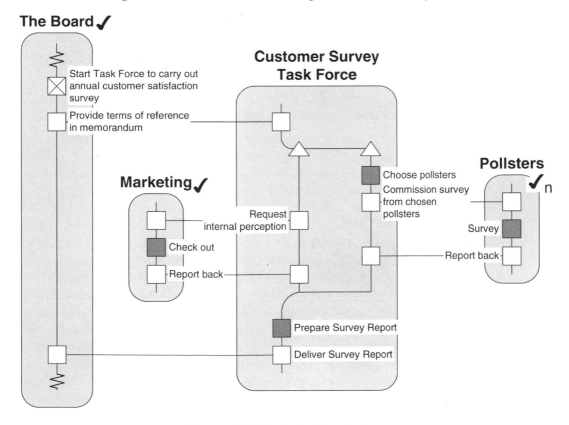

FIGURE 2.39 *A asks B who asks C …*

out the annual customer survey, we will see an interaction between the two roles across which *The Board* gives the *Customer Survey Task Force* its terms of reference. In its turn the *Customer Survey Task Force* delegates parts of that responsibility to other roles: *Pollsters* and *Marketing*, say. The task is broken into smaller sub-tasks and delegated out to other roles. Note that *Customer Survey Task Force* is a transient role, an instance of which is created for the occasion.

As each role completes its sub-task, it might (or might not) report back to the delegating role to say 'I've done what you asked.' So we can expect to find a corresponding 'closure' interaction for each delegation interaction. Or, at least, it is a useful modelling discipline when you see an interaction that represents delegation to look for a corresponding closure interaction in the real process; if there isn't one there you can question whether there should be; and if there is, you can question that too. Looking back at the sample (and not very sensible) process in Figure 2.2, we can see quite readily that the *Divisional Director*, having delegated the execution of the project to the *Project Manager*, apparently never expects to hear about the completion of the project – let's hope that in that model we simply weren't interested in the involvement of the *Divisional Director* after they had started the project off!

In some cases, one role instance might delegate a task to some other role instance which is pre-existent, such as a department or a post in the organization:

- Each March, *The Board* delegates the annual assessment of effectiveness of the company's IT systems to the *IS Department*.
- After an accident in the plant, the *Divisional Director* delegates a review of plant safety to the *Health and Safety Manager*.

In other cases, a role might be instantiated for the purpose: we tend to call such roles 'task forces' or 'project teams'. Their job is to carry out the task and then disband. Figure 2.39 shows just such a role: *Customer Survey Task Force*.

Service interactions

Let's take this a bit further. We have seen how an interaction can be said to 'align' the states of participating role instances: that is, the participating role instances 'go through' the interaction together. As far as the RAD is concerned the interaction is atomic; once it has started we know nothing except when it has finished. But of course, many things happen between parties without apparently any need to synchronize so explicitly. For instance, some roles provide some form of on-demand service: a Line Manager will authorize leave requests at any time, rather than only at some prescribed point in *their* work. So some part of the Line Manager's work must involve 'being available' in case such a demand arises.

This situation is modelled quite straightforwardly in a RAD by having a separate (i.e. concurrent) thread of activity in the service role 'hanging free' – see Figure 2.40. By definition, the activating condition for such a part-

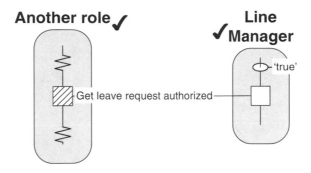

FIGURE 2.40 *A role ready to have an interaction at any time*

interaction on the server's side is 'true', i.e. the role is ready to undertake such an interaction on demand, at any time. Indeed, that thread can be activated/instantiated an indefinite number of times as requests for service come in, since the activating condition remains 'true'. This is often what we want. Not all activity is strictly sequenced in the sense of 'When X has finished do Y'; often the logic is more catch-all: 'Whenever necessary do Y'. The interaction with a hanging thread is a special case of this.

I like to describe free-hanging threads such as this as corresponding to parts of the role's brain: a bit of the brain that is always ready to have that interaction and start that thread.

Primed service interactions

In some situations we might wish to have an activity thread hanging free, ready to respond, but only after some prior action has been completed. For instance, once a business has acquired its tax registration it can process any number of orders arriving asynchronously. We would show this situation in a RAD as in Figure 2.41. When (the single instance of) *A Company* starts, the only thing it can do is obtain tax registration. The *Obtain tax registration* action takes place and the role then enters a state where it is able to accept an interaction (a purchase order). That interaction causes two threads to start: one to process the order and another to wait for another interaction (another purchase order). While the first order is being processed, a second can arrive which again causes the two threads to start: one to process the second order and another to wait for the third, and so on.

We can see this more easily by looking at it in terms of tokens. Figure 2.42 shows the states through which the RAD fragment passes (note the direction of the broad arrow at the back of the diagram):

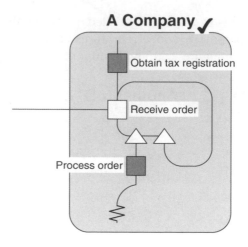

FIGURE 2.41 *Many-at-a-time processing after initialization*

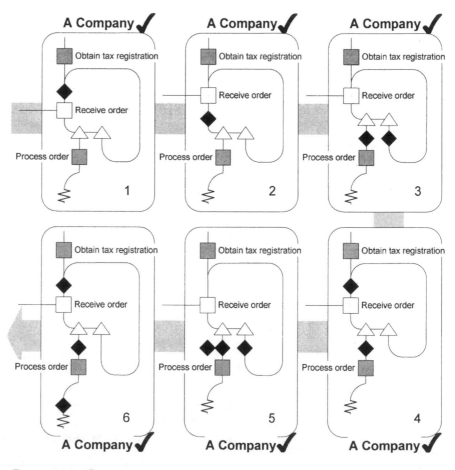

FIGURE 2.42 *The successive states of the company after receiving tax registration*

1. Registration for tax has been completed; waiting for the first order.
2. The first order has arrived via the interaction.
3. An equivalent state to state 2, using the definition of a part refinement.
4. An equivalent state to state 3, remembering that a state line shows a single state. The role is now ready to process the first order and also ready to accept a second.
5. A second order arrives but the first still hasn't been processed.
6. One order (it could be either) has been processed, the other is waiting for processing, and a new order is also being awaited.

The thread to process an order is successively started as each order arrives. In fact, strictly, the actions on that thread are instantiated as and when their activating condition becomes true. We can imagine that the actions on the thread are being instantiated as many times as there are orders. Certainly each 'thread instance' proceeds independently, at its own pace. There is no implication that orders 'queue' to follow the thread. In fact, it is quite possible for orders to be processed at different speeds – the RAD says nothing about which will finish first. I have drawn one possible sequence of markings for the RAD, principally to show how a mass of concurrent activity can build up, indicated by the proliferation of tokens. RADs are about the concurrent activity in the real world.

Strictly sequenced service

In some cases it might not be desirable to have an activity thread hanging free in such a way that it can be activated at any time or indefinitely many times: we might wish to handle requests just one at a time in strict sequence. In this case a hanging interaction is clearly not what is wanted, and in order to serialize the processing of requests we put that processing in a sequential loop. The processing can no longer take place asynchronously: it can only occur when the server has finished serving the previous request. This situation is shown in Figure 2.43. We must assume that when the role starts there is a token sitting on the loop, as we have shown. Spend

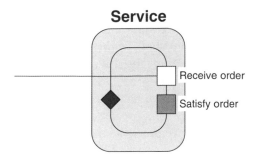

FIGURE 2.43 *One-at-a-time processing*

a moment checking that orders are dealt with strictly in rotation and a new one cannot be received until the previous one has been satisfied.

Conditional interactions

It is not uncommon for an interaction to take place only under certain conditions. For instance, suppose that you (a buyer) order some goods from me (a seller). When you receive the goods from me you check that they are OK, and if they are not, you send them back to me. That interaction for returning the goods to me takes place only if they are not OK. How should we model this? Our first thought might be to draw a process as shown in Figure 2.44. Follow it through, especially on the seller's side.

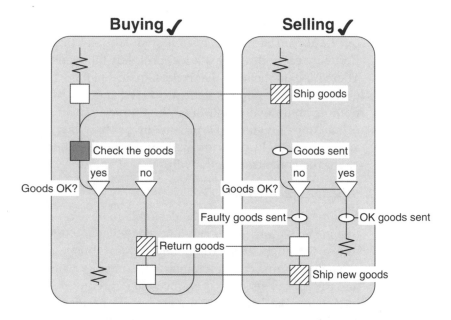

FIGURE 2.44 *A conditional interaction wrongly modelled*

This would be wrong. Firstly, remember that a case refinement does not involve any action: it simply divides the state. So, the figure says that in the *Selling* role, immediately after the goods have been shipped, we must be able to refine the state marked *Goods sent* either into the state *OK goods sent* or into the state *Faulty goods sent*. But if this really were the case it would mean that the seller must know immediately after shipment whether the goods are faulty or not and can be ready to respond appropriately! If we assume that the seller is not so underhand, we have modelled the wrong process. The problem is that the condition *Goods OK?* cannot be determined at that point within the body of the *Selling* role. Only the *Buying* role has the resources within its body to determine the value of the condition.

A correct model for this situation would therefore be that shown in Figure 2.45.

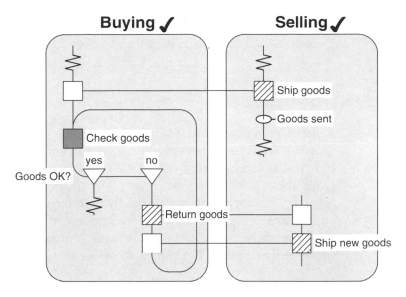

FIGURE 2.45 *A conditional interaction correctly modelled*

Here, we show the case refinement only in the *Buying* role and a second interaction with the *Selling* role that only takes place if the *Buying* role decides the goods are not OK. The *Selling* end of that interaction takes the form of a 'hanging' thread: it can take place whenever necessary and we do not need to tie it into the rest of the role. Only if the *Buying* role takes the initiative and 'forces' the interaction, will the *Selling* role need to do anything about it.

If the seller had a standard practice of always checking with the buyer that the goods delivered were satisfactory then we would explicitly model this of course.

KEY POINTS

Interactions show the points where roles cooperate.

An interaction can involve any number of roles.

An interaction synchronizes the states of all the participant role instances.

The result of an interaction can be the exchange of something, or joint activity, or both.

Replicated part-interactions allow us to model interactions involving all current instances of a role.

Interactions can be modelled in concrete or abstract terms, or both.

Interactions can be more complex than we think and the right level of detail must be struck.

Pupil:	We've spent a lot of time on interactions.
Tutor:	Yes, but don't sound so surprised: processes are about collaboration, and interactions are where collaboration takes place. If you're used to swim-lanes, you might suppose that when two swim-lanes interact, all that happens is that the locus of activity moves from one to another. In fact, interactions are very rich in nature and we must be aware of that.

REPRESENTING TRIGGERS

Sometimes a role must wait for something to happen before it can proceed. To be more precise, sometimes a *thread* must wait for something to happen before it can proceed. We call that something a *trigger*.

We show a trigger by an arrow (➤➤) placed on the state line – see Figure 2.1 – with a caption that briefly describes the trigger concerned. In strict terms, a trigger moves the role concerned from the state preceding the little arrow to the state just after it – imagine this in terms of the movement of a token. In Figure 2.46 we see the start of a thread that waits

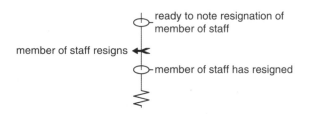

FIGURE 2.46 *The before and after states of a trigger*

for someone to resign. We have labelled the before and after states to emphasize the way the trigger changes the state.

In Figure 2.2 the RAD tells us that a *Divisional Director* role instance must wait until a new project has been approved – somewhere outside this process – before it can proceed.

Triggers marking calendar time and clock time

End of month might be the absolute time that triggers the start of the work to prepare the month-end paperwork (Figure 2.47). *Thursday 6.00 pm*

FIGURE 2.47 *A trigger marking calendar or clock time*

might signal the start of work on the weekly wages cycle. *1 May* might signal the start of the annual budget round.

A trigger of this sort is commonly found at the start of such time-related threads. But it could fall in the middle of a thread. Suppose we issue an invoice and then wait until the end of the following calendar month before checking that payment has been received. This would look like Figure 2.48.

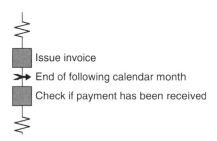

FIGURE 2.48 *Waiting for a time trigger in the middle of a thread*

Triggers marking the passage of time

Sometimes work has to be suspended until a certain period of time has passed. Our little arrow symbol allows us to represent this. *Thirty days later* might be such a 'relative' time trigger: thirty days after an invoice has been sent out we might check that payment has been received. Figure 2.49

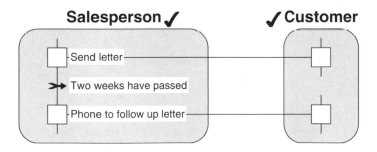

FIGURE 2.49 *A trigger marking the passage of time*

shows a situation where a letter is sent to a customer and the role waits two weeks before making a phone call to follow up the letter.

Triggers marking external events

Sometimes, something happens outside our process which has an impact on its progress – we call this an *external event*. For instance, suppose we are modelling the recruitment process in a company. The resignation of a member of staff triggers a thread of activity in the *Human Resources* role in our process, preparing job specifications, contacting recruitment agencies, and so on. The person resigning would communicate their resignation to the *Human Resources* role in some way. We might not be concerned in our model to know the detail of the resignation, how it was communicated, or any details other than that it has occurred and that a certain post is to become vacant. We would represent this using something like Figure 2.50.

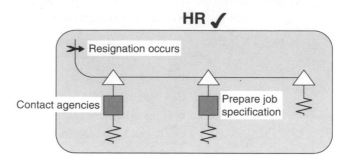

FIGURE 2.50 *An event external to the modelled process*

We could on the other hand adopt the perspective that the resignation is in some way an interaction between the *HR* role in our process and some other role, and we would have something like Figure 2.51. But this adds

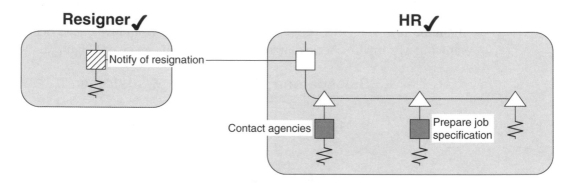

FIGURE 2.51 *Treating an external event as an interaction*

little and only moves the boundary of our model out to some portion of the role *Resigner*, leading us firstly to think (perhaps unnecessarily) about the identity of that external role and secondly, what led up to that interaction in the *Resigner* role! As it stands, Figure 2.51 shows a process which depends on the role *Resigner* deciding to use the *Notify resignation* interaction for anything to happen.

However, if we are concerned, for instance, to check the way we deal with someone who has resigned, it is obviously important that this interaction should appear and, indeed, we might decide we want to capture precisely *how* that interaction can take place formally.

Triggers marking internal events

An external event spots something happening outside the modelled process that triggers action inside the modelled process. We can also use an event to spot something happening 'over there' in the modelled process that triggers action 'over here'. Figure 2.52 shows an example. In this case,

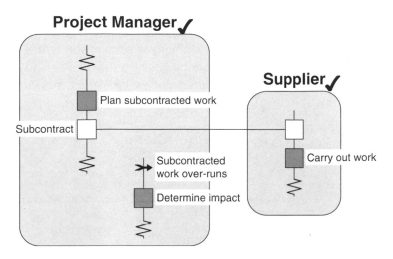

FIGURE 2.52 *Spotting an event internal to the process*

an action has taken more than an allotted time: the subcontractor is late and this triggers the Project Manager into action.

We need to be a little careful when using this way of spotting something happening 'elsewhere' in the process. When you do it, ask yourself the question 'How would this role find out that this event has happened?' If the answer is 'by asking so-and-so', then we might question why we did not model that interaction. The example in Figure 2.52 is fine: the Project Manager has a calendar and a note of when the work should have been finished. The trigger doesn't tell us *how* the Project Manager spotted the over-run, simply that it was spotted.

Triggers marking as-and-when events

Finally, there is the situation where a thread of activity can be triggered at any time. It might be on a whim: 'Let's carry out an audit of the retail side of the business.' An example is shown in Figure 2.53. Indeed, if the thread

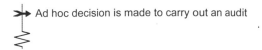

FIGURE 2.53 *An as-and-when event*

can be started on a completely ad hoc basis then we have no need to show a trigger at all.

Triggers can perform magic

An external event trigger is a way of defining part of the boundary of our model. We are recognizing that something happens 'out there' that affects 'in here'. The actual effect can include a little magic: the triggering event might cause the importing of something into the process (strictly, into the role body) or it might change the state of something already in the process (strictly, inside the role body).

Figure 2.54 shows an event – *Application form arrives from customer* – that imports something into the role: the application form from the

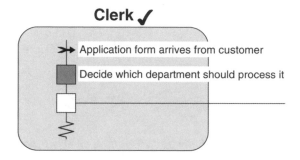

FIGURE 2.54 *An event importing something into the role body*

customer. The fact that we have shown its arrival as an external event says that we don't care, for the purposes of this model at least, how this happens. It just does. The *Clerk* role now has that application form in its possession, in its role body.

Figure 2.55 shows an external event changing the state of something already in the role body: the sales predictions in the possession of the Chief Accountant have, somehow (and we're not saying how), been changed.

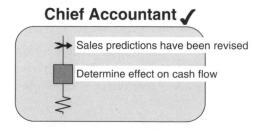

FIGURE 2.55 *An event changing the state of something in the role body*

Again, when we use a trigger in this way, we should ask how the prop was changed: perhaps it was through an interaction with some other role. Should we be showing that interaction on this model?

Choosing and modelling triggers

Triggers and threads

This is an appropriate moment to remind ourselves that a role may have more than one thread. Figure 2.56 shows a part of a *Divisional Director* role

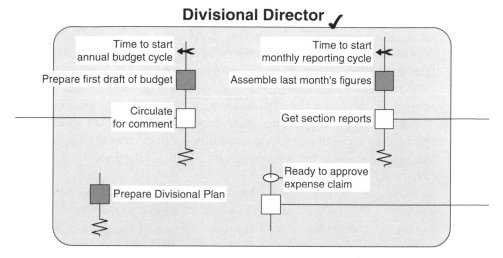

FIGURE 2.56 *A role instance with four threads ready to start*

which has four threads: one for dealing with the annual budget cycle; one to deal with the monthly reporting cycle; one waiting for an interaction with anyone wanting an expense claim approved; another to get straight on with preparing their Divisional Plan. We can think of these threads as corresponding to four parts of the Divisional Director's brain: each waits for its own particular trigger to set it off.

Event-driven vs cycle-driven threads

An *event-driven* thread is one that is triggered whenever a certain event occurs. It is a thread which represents our response to a certain situation. We can imagine many different sorts of triggering events. For instance, *Customer application arrives* and *Sample arrives for analysis* might be the triggers for threads that deal with individual cases: we will have a number of customer applications being processed at any one time, and there will be a number of samples at various stages of analysis at any given moment. *Budget change announced*, however, would be more likely to trigger a process that starts, proceeds and stops ... and maybe is triggered again at some time in the future on the next budget change; but we are not going to be dealing with several simultaneously. In the worst case, we'll abandon any (over-running) attempt to deal with the last one and switch our attention to the one that has just come in.

A *cycle-driven thread* is an event-driven thread which fires regularly according to the clock or the calendar. There is probably (but not certainly) only one instance of such a thread in progress at any one moment and we fire an instance off at regular intervals. We will find events such as *End of Month*, *5 April* and *Midnight* at the beginning of such threads.

Plans and activities

It is not uncommon for certain parts of a process to be started off according to the dictates of a plan which is prepared *during* the process we are modelling. In other words, the *threads* of activity that will be carried out are known, but exactly which threads will be done and in which order they will be done might be decided on-the-fly, as the process proceeds. Those decisions – what and when – are what plans are all about. Our plan will tell us that when a certain condition is right, a certain thread of activity should be started, but we cannot tell which or when until we are into the process.

Suppose our business is developing new electrical goods. At various points in the design and development of a new product we shall need to carry out various tests and obtain certificates of compliance with certain regulations. Precisely which tests are needed will vary from product to product, and they will be different for a toaster and a hand-dryer. To model such a general **Develop a New Electrical Product** process we might therefore need to show threads of activity to do with carrying out tests and obtaining certificates, with those threads being triggered by internal events: effectively 'The plan says the moment is right.' Figure 2.57 gives an example. We have gone a step further and modelled the (internal) event that all the planned activities have been completed.

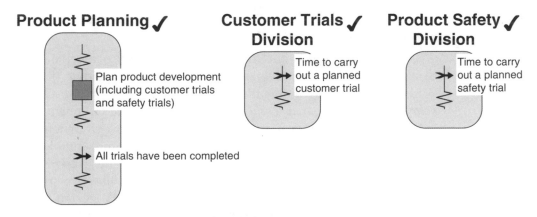

FIGURE 2.57 *Modelling a plan and its activities*

Modelling exception conditions – timing out an action

Suppose a Helpdesk takes calls from customers. If the Helpdesk person cannot answer the question there and then on the telephone, they tell the caller that they will go away and investigate the enquiry and call back with an answer within the hour. If at the end of the hour they don't have an answer to give the caller, they need to give an interim report on things and perhaps give a time by which they will get back. We want to show both the intended process – immediate answer or call back within an hour – but also the process if an hour passes without an answer emerging from the investigation.

To do this, we show the handling of a late action as a response to the event which is the timing-out of the action. And when the action does finish (or gets aborted perhaps) we take an appropriate course of action depending on whether it has timed out – see Figure 2.58. We are using the

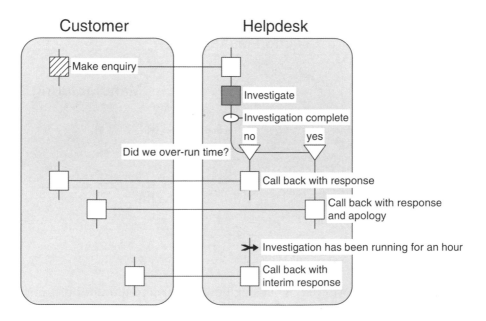

FIGURE 2.58 *An action runs late*

➤➤ to detect an appropriate event *inside* the process: *Investigation has been running for an hour*. This trigger sits on its own thread – it doesn't need to be 'tied into' other parts of the process.

When people model processes, they all too easily fall into the trap of modelling the normal or expected behaviour. But if we observe people doing a process we shall often find them dealing with abnormal or unexpected situations. Should those behaviours be part of the model too? Well, the answer depends on the purpose of the model. But we might say that if we want a full picture of an existing process, or a thorough model of a planned process, then we had better not forget those exception conditions. They will often reveal problems elsewhere in the process: earlier work that has not been completed on time, poor quality inputs such as forms inadequately completed, perhaps because they are too complicated, and so on.

Modelling exception conditions – timing out an interaction

Once a customer has placed an order, they wait for the interaction which is the delivery of the ordered goods. Do they wait for ever? Of course not: if the goods have not turned up after a certain period they will want to chase them. So, how would we model this exception condition and the way it is handled?

Suppose a Section Manager waits for goods from Manufacturing and for the corresponding paperwork from QA before passing the goods with the paperwork to Shipping for shipping. An optimistic view of this process is shown in Figure 2.59.

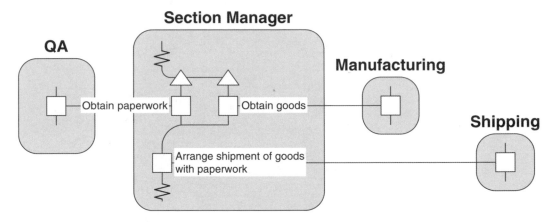

FIGURE 2.59 *An optimistic process*

If we were modelling the process and did not press our questions very hard, this is the story we might hear. But of course we might guess that the Section Manager won't wait for ever for the paperwork to turn up. Further questioning might reveal that the process as actually implemented takes a rather more pessimistic view of the world and has a workaround for when the paperwork is late, in other words, for when the *Obtain paperwork* interaction times out as perceived by the *Section Manager*.

That pessimistic – but realistic – process is shown in Figure 2.60. The *Section Manager* starts the two interactions with *QA* and *Manufacturing* as before but now, having received the goods from Manufacturing, has two possible courses of action: one in the case where the paperwork arrives on time (on the other thread of the part refinement) and another in the case where that interaction with *QA* times out. In the first case, the *Section Manager* proceeds as normal, shipping the goods with the paperwork. In

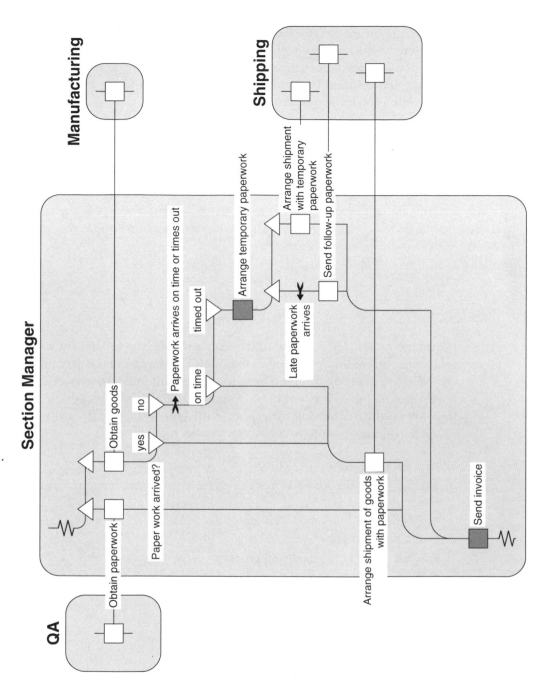

FIGURE 2.60 *A pessimistic yet realistic process*

the second case, they ship the goods with temporary paperwork and, at the same time, wait (even longer, in fact indefinitely according to the RAD) for the real paperwork to turn up. When it does, the follow-up paperwork can be sent off and the process resumes its 'normal' course, with an invoice being sent to the consignee.

More complicated mechanisms can be modelled using a ➤➤ to pick up timeouts and respond to them. Spend a moment checking how the two triggers are used in Figure 2.60.

Triggers can be complex ... in their captions

Figure 2.61 shows some typical situations modelled in different ways. The first RAD fragment shows a role that has to prepare a report at the end of

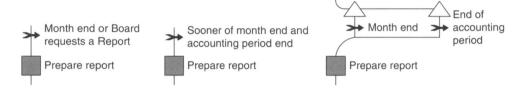

FIGURE 2.61 *Some tricky triggers*

the month or whenever the Board requests a report. The second shows the role writing the report at the end of the month or the end of the accounting period, whichever comes first. These make the point that an event can be quite complex but only needs to be described in English in its caption. The third shows the role writing the report at the end of the month or the end of the accounting period, whichever is the later.

> **KEY POINTS**
>
> The trigger symbol is used to represent a variety of things that can trigger activity in a role:
> - moments in time;
> - the passage of time;
> - events outside the modelled process;
> - events somewhere inside the modelled process;
> - as-and-when events.

REPRESENTING THE AD HOC PROCESS

Pupil: So far we seem to have been describing processes that have some order, some sequence to them. OK, we've seen roles with several threads – separate parts of their brain for different parts of their responsibility, as you put it. But some processes are entirely ... ad hoc. Things don't happen in a predefined order. Surely RADs aren't appropriate for them?

Tutor: Au contraire! You've just said how they can be handled: with separate threads! Yes, we can imagine a process where the order in which activities

Pupil:

are carried out is decided by the actor on-the-fly, and can't be laid out beforehand. Writing a book is one: I flit between writing a chapter, changing a style across all the chapters, amending the index, updating the contents list, and so on. A part of my brain is ready to do any one of those things at any time. There can be many such parts, each corresponding to a thread of spontaneous activity, perhaps a single activity. With that thought, could you now sketch the RAD of the process I just described?

Pupil: Well, each of those threads must sit on its own ... I'm not sure if they have triggers on them ... I guess I don't need to show triggers if they are truly spontaneous as far as the model is concerned. How about this (Figure 2.62)?

Tutor: Looks fine. Here's a role with six bits of brain each ready to start a thread when the actor feels like it – pure ad hoccery!

Author

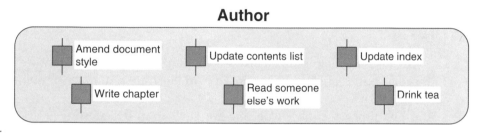

FIGURE 2.62 *A role of ad hoc activity*

REPRESENTING PROPS

Processes involve things. The states of things affect the course of events and are affected by events. When I work on a chapter of the book, the state of that chapter changes. When the chapter is finished I can have it reviewed.

These things are the props for the role. Props can include the resources the role needs.

We might be happy to simply note the props concerned in captions to actions, interactions and decisions:

- Prepare the Project Plan.
- Assemble the engine.
- Deliver the book that was ordered.
- Hand over the terms of reference for the work.
- Is the purchase over £100?
- Was the employee working abroad?

But where are these things – the Project Plan, the engine, the terms of reference? In the words of Bluebottle: 'Everyone's got to be somewhere.' The Riva answer to this is 'In the body of a role instance'. If I am preparing the Project Plan as (an instance of) Project Manager, the Project Plan is on my desk. If we, an Engine Construction Cell, assemble the engine, it is in our assembly area. If I am deciding if the employee was working abroad, something in my role body must allow me to decide. And that something must have got there somehow.

This is more than a matter of geography or cubicle politics of course: my desk is where I keep my resources to carry out my role; our assembly area is where everything we need to assemble engines – equipment, tools, parts – is held for our job. When parts are delivered (via an interaction) they end up in our role body. When we deliver an assembled engine, it passes from our role body to that of the Final Assembly Team.

If it's appropriate for the model in hand, it makes sense therefore to show props within the grey area of the role on a RAD, just as a simple list.

CASE STUDY 1

To close the chapter, let's take a look at a RAD of the sort that we might produce in real life. I have chosen it to illustrate a number of points. See Figure 2.63.

The process is of a type that we shall soon be referring to as a case process: it's a process that deals with one something, in this case a 'candidate product'. So the process is triggered when the Marketing Director detects the fact that a candidate product has been identified. A new responsibility is created to deal with it in the form of a Product Manager (the only role with no pre-existing instances). There is then a flurry of concurrent activity from a number of roles in the preparation of a dossier that goes to the Board. They might reject it (end of process), or ask for it to be resubmitted, or pass it for development. During development, the Board is kept up to date with progress, and finally Product Assurance run an Acceptance Test over the product before it is deemed ready for sale. The process therefore has two possible outcomes.

What is noticeable about this model is how the interactions outnumber the actions. It was drawn to emphasize the collaborative nature of this process, to stress how everyone's involvement has to be coordinated to ensure success. Since time-to-market is important for a product company, it also aimed to show where different roles would be operating concurrently and there are part refinements within roles, allowing them to get on with several things at once. At one point, Development can be involved in at least three concurrent threads of work. Once again, interactions are important in getting material out to people and gathering it back in. Coordination.

In later chapters, we shall look at how a RAD can be used to get to answers to the sorts of questions we might ask about a process. But as well as providing us with an example of a RAD, perhaps we can also look at the sorts of things we can spot when something like this is drawn:

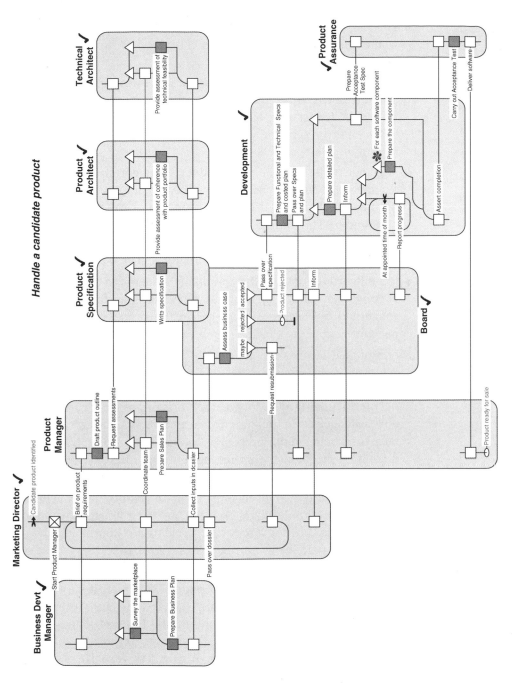

FIGURE 2.63 *Handle a candidate product*

- The use of grey boxes to contain the activity of roles helps us see where roles fit in the process (much better than swim-lanes, we might say).

- We might quickly observe that some key roles are involved early on, but disappear from the work in the later stages. The Marketing Director, for instance, has no involvement apparently once a project has been given the go-ahead by the Board – was this really the case, or just appropriate for this model?

- A glance tells us that the Product Manager moves into a rather responsive style of management once the project has been passed to the Board for approval and – if approved – on to Development for production. They appear to be on the receiving end of interactions from others. What are they doing to manage things proactively? Development seem to be in the driving seat.

- If the Board does reject the candidate, it seems to be a private matter: no one else gets to hear about it. Perhaps as far as this model is concerned, we weren't interested in the process for rounding off a rejection.

- Acceptance Testing can, it appears, only result in success. Do we ever get to the point where we decide that it has all been a terrible mistake and abandon the thing in Acceptance Testing? Indeed, we can imagine the product being abandoned at all sorts of points: if costs spiral or the timescale goes out too far, for example. We haven't shown those measures being monitored or responded to. This could be because there is no mechanism for that monitoring or because we chose not to model it.

- We presumably could have shown Development developing the product as a simple black box, but we have chosen to unpick it at least as far as showing a reporting cycle to the Board, and the need to develop in parallel a number of components that make up the product.

- The model is largely abstract in nature. Although it shows real posts and groups in the organization, the actions and interactions are generally expressed in terms of intent: the Product Manager 'requests assessments' and we are not told how that request is made or what form the assessments take (written, emailed, verbal?).

These few observations should make it clear that as well as allowing us to capture the dynamics of a process precisely, a RAD is 'revealing': it allows us to see the nature, the style, the flavour of what is going on, and hence allows us to answer the sort of questions we shall want to ask about the process.

3 Dynamism in the process

Highlights the levels of within-process concurrency that can be captured in a RAD.

INTRODUCTION

At the end of Chapter 1 our Tutor pointed out two important messages: instantiation and concurrency. Things happen because many concurrent things happen at the same time. Remember the chevron of geese. Think of a beehive. When we draw a RAD we are essentially drawing a static model that shows the relationships between types of things: types of roles, types of actions, and types of interactions. But more importantly, it describes the potential *behaviour* of an organization when it carries out the process that we have modelled. That behaviour arises from instantiation. In this chapter we shall look more closely at how a RAD captures concurrency. We shall do this by showing how we can pick up a RAD and 'run' it – a sort of paper animation. Such a paper exercise is useful for understanding or validating a process model; if we want to enact the process on a BPMS then the topic becomes central as we shall see in Chapter 13.

As we looked at each concept in the RAD notation in Chapter 2 – case refinement, part refinement etc – we found it useful to animate the fragments of RAD in order to understand the sorts of behaviour that we had effectively defined. In this chapter we shall look at the business of animation on a larger scale: that of the entire RAD. How do we look at a RAD and animate it to see what the process *does*?

A RAD describes a process in terms of the relationships between *types* of roles, types of actions, types of interaction, and types of events. When we animate a RAD to see how it works, we look at *instances* of roles, actions, interactions and events, and we are interested in the actual *states* of role instances.

A REMINDER ABOUT STATE

Each line (other than those indicating interactions) represents a potential state of a role instance. Because a role instance can have more than one thread of work active at any one moment, we can be stricter and say that each line represents a potential *sub-state* of a role instance. For instance, when a role instance enters a part refinement with three threads, we can think of its state becoming the 'sum' of the sub-states on each of those three threads. We showed the state of an instance by placing tokens on the

117

appropriate state lines. As the role instance does its work – carrying out actions, taking part in interactions, responding to external events – so the positions of the tokens change: the marking changes to reflect the change of state of the role instance. We can imagine a software tool on our PC that shows us our role instance – RAD-style – with tokens on the state lines to tell us where we have got to in carrying out that role. As we do items of work, so the marking changes to reflect the way we are moving through the process.

If we stood back and looked at a running process we would see a set of role instances, each in its current state. If we asked the question 'What is the state of the process?', we could answer by saying that it is the 'sum' of the states of all those separate role instances. This matches real life: 'How far have we got with dealing with that insurance claim?', 'Well, the loss adjuster is currently waiting to arrange a meeting at the claimant's house, the clerk dealing with it has checked with the police, and we're sorting out liability with the lawyers right now.'

We will be able to tell when the process has reached its goal when the desired states are reached in certain of the role instances.

HOW ROLE INSTANCES 'START'

Let's remember that there are two ways for an instance of a role to arise. The simplest situation is where a role has one or more pre-existing instances. Those instances are in place when the process starts and can begin work. Other roles need to be instantiated once the process is running. Suppose that a process has just started and there is a pre-existing role instance for one of the roles, or a role has just been instantiated in a running process (e.g. a Task Force has been set up). What happens to that role instance? To answer this, let's step back for a moment.

Inside a role we draw all the things a role does: actions and part-interactions. For each of these we can define the activating condition: the state which allows the action or part-interaction to start. We can define those states by making the post-condition of one thing the activation of the next: by connecting them with a line. So in Figure 3.1 when we have done *A* we can do *B*: or in our jargon, the post-condition of A is the activating

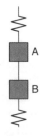

Figure 3.1 *'After doing A we can do B'*

condition of *B*. (Strictly, once the instance of *A* has completed, an instance of B is created and can be acted by the current actor of the role instance.)

Suppose now that we have a thread of activity whose beginning is as shown in Figure 3.2. What is the activating condition of action *A*? The rule

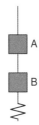

FIGURE 3.2 *The start of a thread in a role on a RAD*

is that at the head of a thread the state is 'true', and since the value of this is always 'true' we can deduce that action *A* is always ready to run. In other words, for any instance of the role there is always a token sitting on the state preceding *A*, as shown in Figure 3.3.

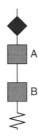

FIGURE 3.3 *A role instance thread ready to start in a running RAD*

This generalises of course to each such thread in the role: when a role instance starts (in particular when it has just been created), it has a token at the start of every separate thread. What happens on each thread then depends on the first thing on the thread:

- It waits in front of an action, in which case it is up to the actor to decide when to start the action.
- It waits to take part in an interaction.
- It waits for an event to happen.

Let's revisit the fragment of a process we saw in Figure 2.56. When the *Divisional Director* role instance in that process starts, there will be a token at the top of each thread, as shown in Figure 3.4. What we observe is the *Divisional Director* waiting for the moment to come when they must start work on the annual budget; waiting for the moment when it is time to start working on the monthly report; waiting for an interaction with someone

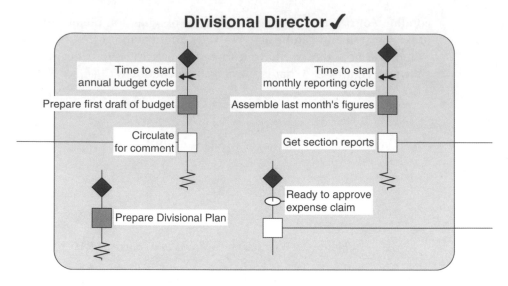

FIGURE 3.4 *A multithreaded role instance ready to start in a running RAD*

wanting an expense claim approved; but able to get straight on with preparing the Divisional Plan. If we restrict ourselves to this small fragment, we have a role instance that is ready to start four threads of activity.

Note how important it becomes when we draw a RAD to mark roles with pre-existing instances, if we really want to understand how much concurrency there is at the outset and hence how the process can unfold.

HOW A PROCESS STARTS AND RUNS

When we look at a RAD we want to be able to see how and where it starts. Simple. We just ask ourselves 'Which roles have pre-existing instances?' Each of those instances then starts in the way we have just described: with a token at the head of each free thread. We need say no more. We simply use the rules we developed in Chapter 2 to see how the tokens flow. When the role instance comes across a part refinement, more threads will become active: the concurrency of the role instance will increase. Some threads will merge and others will come to a dead end. If the role instance comes to a replicated part refinement, a multitude of concurrent threads can be started up.

This is a good moment for a caution. When you draw your first RADs you will find yourself unable to resist two things: joining everything up, and creating unnecessary sequences. If you succumb to either of these temptations you will deny the existence of concurrency. For instance, do those four black-box actions you have drawn in a sequence actually have to be done in sequence, or is some parallelism possible?

HOW ROLE INSTANCES 'END'

We have no special notation in Riva to mark the 'death' of a role instance in a RAD, no symbol that says 'delete this role instance'. When we want to show this we simply use a black-box action with an appropriate caption: e.g. *Close down the Task Force, Close off responsibility for Project Managing*, or *End role of Expense Claimant*. If this troubles you, remember that a role's activity can be made up of many concurrent threads. It is not always a matter of reaching the end of a single thread.

HOW A PROCESS 'ENDS'

A process ends when there are no role instances with something to do ... assuming of course that they cannot be revived into activity by an external event that they are waiting on.

Later, next to the water cooler

Tutor:	I hope you can see now why a RAD is not a flowchart.
Pupil:	Well, a flowchart – and, I guess, all the variations on the theme that I've seen being used – represents a simple, single, sequential flow of activity, a single thread with no possibility of multiple concurrent activity. A swim-lane diagram typically does no more than show the single thread trundling between roles. Even if we allow a thread to divide we aren't getting the true concurrency that happens in the world.
Tutor:	Exactly. It is all too easy to model a process as if it is some kind of simple sequence – processes are hardly ever like that. If they were, most people in the building would be standing still at any one moment, waiting for their turn to come. Instead, as a process unfolds, there is a constantly changing flux of instances that ebbs and flows. We cannot think of a process as a sequence of activities, some beads strung on a string, pieces of meat on a kebab stick. We must accept that it is more like ... well, like the workings of an organization! So although a RAD is a static model that shows relationships between types of things (roles, actions, interactions), it actually captures the potential dynamics; what happens when things run and instances happen.
	Let's look at the RAD in Figure 3.5. Tell me about the concurrency in it.
Pupil:	OK. When the process starts there is a Project Manager, a Board, an Accounts Department, a Client, and a Line Manager. The model seems to be mostly about the Project Manager, who has six separate threads. One thread is triggered at the end of a quarter and they then have a quarterly review with the Board. Another triggers every two weeks to have an interim progress review with the Line Manager. Then there is another at the end of the month which is to do with getting the monthly report written and sent off to the client.

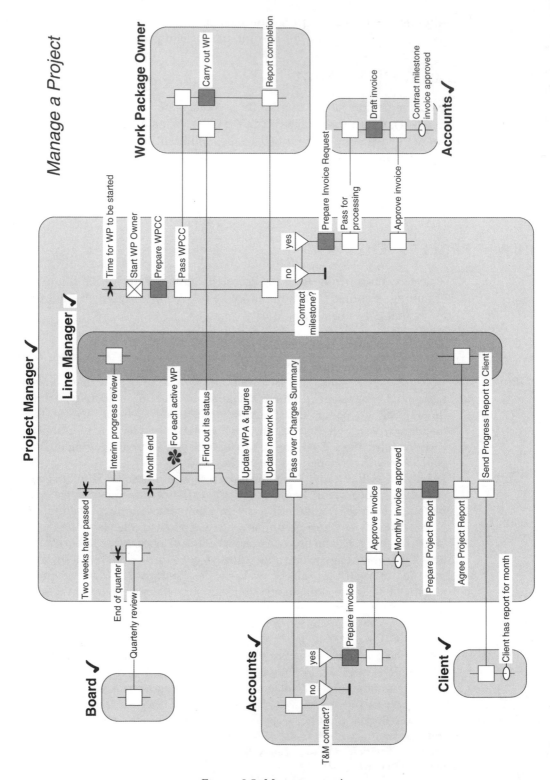

FIGURE 3.5 *Manage a project*

The most interesting seems to be one to do with handing out pieces of work – 'work packages'. Each work package has a responsibility associated with it in the form of a *Work Package Owner* role. When one of those owners is started they get a WPCC – some sort of terms of reference I guess – and they do the work – no details about that – and finally tell the Project Manager when it's finished – the Project Manager waits to hear about that.

Tutor: So tell me about the concurrency in this process.

Pupil: That work-package-related thread can fire as many times as necessary so, in principle, at any one moment the Project Manager could be having a quarterly review with the Board, having an interim progress review with the Line Manager, preparing the monthly report – which itself could involve as many threads as there are active work packages – and looking after as many threads again for the work packages themselves. So if there are N current work packages, the Project Manager could have $2 + 2N$ concurrent threads of work in hand – that sounds like project management!

And of course there are N instances of *Work Package Owner*, each of which has two threads going: one doing the work package and the other ready to report on status to the Project Manager.

Tutor: Right. But this amount of concurrency is still not enough. So far, we have only talked about the concurrency possible in one process. We now have to step back and get a handle on how the collection of different processes work together and we shall then add whole new levels of concurrency.

KEY POINTS

When an individual process (instance) runs there is concurrent activity. Each role type can have zero, or one or more independently active instances at any one moment.

Within each role instance there can be zero, or one or more threads of activity operating independently and concurrently at any one moment.

4 Process relationships

Examines the two types of dynamic relationship that processes can have and how to represent them in RADs.

WHAT HAPPENS IN AN ORGANIZATION?

So far, we have seen how a Riva RAD can be used to capture the concurrent and collaborative activity that makes up a single business process, in terms of roles and their actions and interactions. But if we walk into a company's building, we might guess that going on around us are many processes dealing with the many different aspects of the organization's life. As we look around, we sense a number of things:

- Some of them are 'big' processes – the development of a new pharmaceutical drug involving thousands of people across the world – and some are 'small' – someone claiming their expenses.

- Some are long-running, and some are over quickly – think again of the two processes in the above example: ten years against ten hours, say.

- Some processes support others: during the process of running a student examination, the process of marking an individual script is carried out many times; during the development of a pharmaceutical drug, many clinical trials are carried out; during a clinical trial, many medical tests are carried out.

- Some processes interact with others: the process for designing a new product interacts with the process for making a test batch; the process for preparing the annual budget interacts with the process for planning marketing campaigns.

We sense that all the activity going on around us in the organization is a *network* of interacting processes, a network that is changing by the moment. Some processes are occurring many times at the same instant: 124 different clinical trials are 'in progress'; 2,489 expense claims are 'going through the system'; 167 papers are 'in the process of' being marked and moderated.

Let's think about that last paragraph more closely. We have one process for running a clinical trial, but there are 124 *instances* of that process in action at this moment. This sounds familiar. Just as it was possible for an action or a role *type* to have many instances at a given moment, a *process type* can have many instances in progress at any moment. And those instances come and **go** in the same way: when the 33 Physics students have sat the examination paper, 33 new instances of the process for marking an exam paper must be created. When the result for an exam paper is finalized, so the corresponding instance disappears, its job done.

(Note that we must be careful now to say whether we are talking about a process type or a process instance. I shall distinguish between them if it is not obvious which is being referred to. So remember that the word 'process' on its own really needs to be read as 'process type' or 'process instance' as the situation demands.)

It's hard to over-emphasize the importance of these points when we look at how we will think about and represent processes in some sort of model, whether we are defining, analysing, designing, or improving processes. Let's see why.

If we fail to note that there are many process instances running at one moment, we will ignore the question of what makes them start, and when and how. We will ignore the problems that arise from managing all that concurrent activity. We will ignore the effect of all that concurrency on productivity, on resources, on scheduling, and more.

When we describe a process in a RAD, we draw a static structure of role types. But we saw how, when that RAD 'runs', instances of roles can be created dynamically and, within the role instances, instances of activities and interactions are created dynamically. The RAD captures the dynamics of the process and all its potential concurrency, by describing the types and how they get instantiated. We need a similar approach to describing an organization in terms of its processes. We can draw a static network of process types. But when that network 'runs' there will be a dynamic network of interacting instances. In Chapter 6 we shall look at how we can determine that Process Architecture Diagram (PAD), that captures the dynamics of the organization and all its potential concurrency.

The important conclusion is that if we want to model activity at the organizational level then we shall need an approach that captures the dynamic relationships, and that captures the network and the way that it operates. We now have some important questions to answer:

- How do we decide what process types an organization has? Put crudely: how do we chunk all that organizational activity? Putting aside which processes are started when and how they interact, what is the list of processes used by the organization?

- Given that we have a network of interacting processes, what sorts of interaction can processes have? What sorts of relationship can exist between two processes? And remember that we are interested in dynamic relationships, not static relationships.

- Knowing what processes the organization has, and the sorts of relationship that can exist between processes in a network, how do we decide precisely what dynamic relationships this organization has between its processes?

If we can answer these questions we should be able to walk into the building and, after some analysis, draw a picture of the network of process types that the organization must have: its *process architecture*.

We shall answer these questions in a different order, and start by looking at the two main types of relationship that can exist between two processes, and how those relationships get modelled in a RAD. They are:

- Interaction: where the two processes operate independently but interact at various points.

- Activation: where one process starts another, which then operates independently.

I must make an important point here. In each case, what we want to do is recognize that we will be drawing a RAD for each process in the relationship. We want each of those RADs to be free-standing and 'readable' on its own. But processes cannot be cleaved apart so cleanly; if we cut an arm off a living body we chop through nerves and blood vessels, and to get a truc picture we need to show where they came from or were heading. So we want a way of producing free-standing RADs whilst still showing where they are related or connected.

Finally, before moving on, I want to stress again how nervous I am about hierarchies and decomposition when we are thinking about processes. In Riva you will never see a hierarchy, you will only see networks.

> **KEY POINTS**
> One process instance can create an instance of another by *activation*.
> Two process instances can collaborate via *interaction*.
> From moment to moment, there is a flux of interacting process instances at work in the organization.
> Organizational activity is the operation of an evolving network of concurrent, interacting process instances.

INTERACTION OF PROCESSES

When we look at an organization we know we will see many processes operating – strictly, many process instances. We also know that these processes do not operate entirely separately from one another; in particular two processes will occasionally interact in some way. For instance, a company might have an annual budget-setting process in which it reviews the portfolio of projects and decides on the budget for each in the coming year. At the same time, each project will follow a project lifetime process that might take several years (i.e. several rounds of budget setting). At points during its lifetime, the project (process) will clearly need to interact with the budget-setting process to find out what its budget is to be. We will probably model the two processes on separate RADs. But we know that the processes interact at various points, so we will want to show those interactions on their respective RADs whilst allowing each RAD to be free-standing.

What precisely do we mean by 'process *A* interacts with process *B*'? In Riva we know that everything happens within roles, so a process interaction will clearly show itself as a role interaction. Firstly, this tells us that there is at least one role that the two processes have in common; in other words, at least one role has a responsibility in each process. Secondly, we can expect to find a state of that role which appears in both processes; such a state represents some sort of synchronization between the two processes. It is as if the person carrying out the common role can say 'When I'm here in this process, I'm there in that process' – the common role has a common state too.

The modelling of process interaction

Let's look at the 'rule' for modelling a process interaction and follow it up with some examples.

When we model an interaction between two processes, *A* and *B*, it is best to start by modelling it as an appropriate role interaction in each process/ RAD and then either to pare things down or to beef things up, as the situation demands.

- Find the point of interaction in terms of the two roles in *A* and *B* that interact. Let's call them *RA* and *RB*.

- Draw that interaction in the RAD for *A* at the appropriate point and in the RAD for *B* at the appropriate point. We can represent the interaction differently in the two models if that is appropriate. Note that *RA* and *RB* now both appear in both RADs.

- On the RAD for *A*, name the pre- and/or post-states of the part-interactions for *RA* and *RB*. Give the same names to the corresponding pre- and/or post-states of the part-interactions on the RAD for *B*.

Figure 4.1 shows the final situation in this general case. In both RADs we have shown the role interaction that constitutes the interaction between the processes. We have then labelled the pre- and post-states of each part-interaction, and used the same state labels in both RADs. You can think of the shared states as solder points that make electrical contact across – create the same potential in – the two RADs. In the RAD for process *A* we show a minimal amount of role *RB*. In the RAD for process *B* we show a minimal amount for role *RA*.

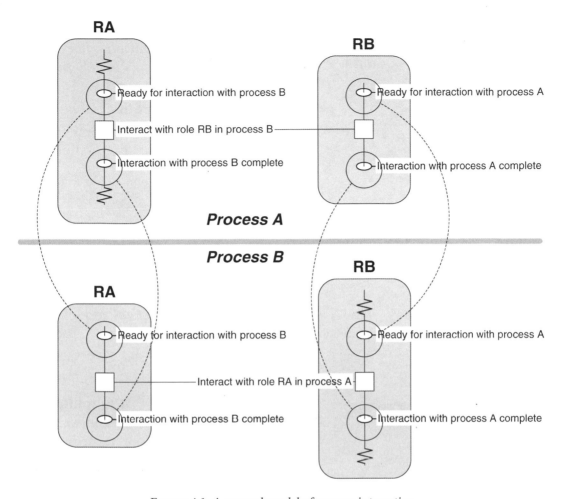

FIGURE 4.1 *A general model of process interaction*

A sample process interaction

Consider the simple example in Figure 4.2. In the **Carry out Annual Portfolio Review** process, the *Product Divisional Manager* is told of a change in their budget as decided by the *Product Portfolio Management Committee*. As far as that process model is concerned we are not interested in how the *Product Divisional Manager* responds to that, simply that they are left in the state *Budget change received*: their response is the subject of the interacting process: **Handle a Product**. The *Product Divisional Manager* plays a part in that, too, and it is in the model of that process that we map their response. Note how the two interacting roles appear in both models, as does their interaction.

The *Product Divisional Manager* and the *Product Portfolio Management Committee* are the two roles that are common to the two processes. Their

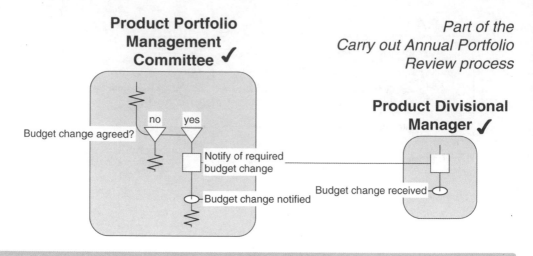

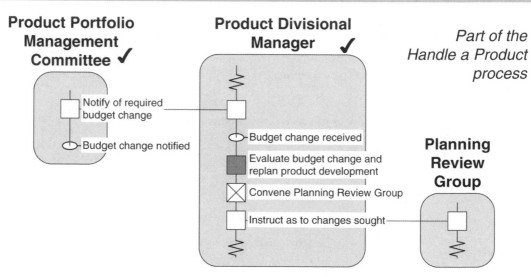

FIGURE 4.2 *A simple process interaction*

respective post-states – *Budget change received* and *Budget change notified* – appear on both RADs.

Figure 4.3 takes a slightly more minimal view of the process interaction as a simple example of the modelling choices open to us. Other variations are possible of course.

In **Carry out Annual Portfolio Review** we have shown the role interaction between the *Product Portfolio Management Committee* and the *Product Divisional Manager* explicitly: *Notify required budget change*. The *Product Divisional Manager* is then in the state *Budget change received* after the interaction.

In **Handle a Product** we have used the action *Take part in annual portfolio review process* to stand for the interaction that the *Product*

Product Portfolio Management Committee ✔

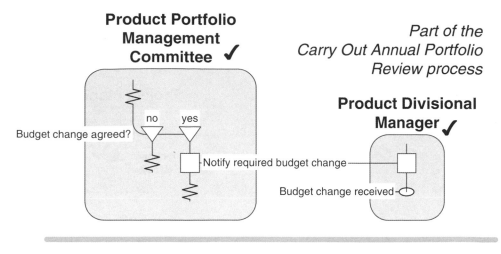

Part of the Carry Out Annual Portfolio Review process

Product Divisional Manager ✔

Product Divisional Manager ✔

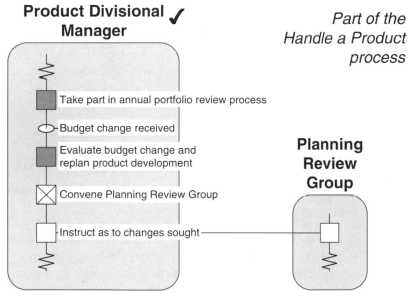

Part of the Handle a Product process

FIGURE 4.3 *Figure 4.2 slightly reduced*

Divisional Manager has in **Carry out Annual Portfolio Review**. The *Product Divisional Manager* is then in the state *Budget change received* after that action.

Clearly, in some instances the interaction could be much more complex with a number of state equivalences defined to tie them together. But note that we always do it by giving the same name to states in the *same* role in the processes that are interacting.

Figure 4.4 shows the same basic modelling scheme at work in the situation where a replicated interaction is involved. In the **Manage a Programme** process, the *Line Manager* goes separately to each *Project Manager* and gets the status of their project. How the *Project Manager*

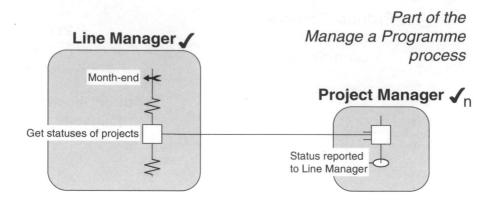

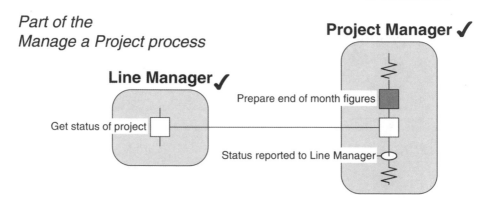

FIGURE 4.4 *Process interaction modelled with two role interactions*

determines that status is the subject of another process altogether: **Manage a Project**. In the RAD for that process we have chosen to show the same interaction, but now at the appropriate point in the management of a project. To tie the two RADs together we have labelled the shared state (*Status reported to Line Manager*) in a shared role (*Project Manager*). We could also have labelled the other three states on either side of the part-interactions in the two roles. That is a modelling decision.

In principle, we can of course cut the boundary between two interacting processes at slightly different places, and there are many ways of representing the boundary. Where we draw the boundary and what level of detail we show depends on what we find useful for our purpose. When we move into the next chapter and start to look at how to chunk organizational activity into processes, we shall find that in practice the boundary is generally clear and the representation straightforward.

KEY POINTS

When two processes interact, at least one role will appear in both their RADs.

The interaction is minimally a common state in the shared role. That shared state represents the synchronization of the two processes.

Whether we show all the shared roles and/or all the shared states and/or all the interactions between the roles is a modelling decision.

ACTIVATION OF PROCESSES

The other important type of relationship between processes is the one where one process is able to 'activate' another, to 'set it going', to 'kick it off' or 'start it up'. The assumption is that if 'process *A* activates process *B*', then process *B* can then go its own way independently of process *A*. The two processes might subsequently want to interact in some way.

As an example, suppose we have a **Carry out a Strategic Review** process in a company. As the mission-critical success factors and future strategic goals are examined during the review, new targets are set for different parts of the company and those targets in turn lead to the need for a number of Tactical Reviews. We can think of the **Carry out a Strategic Review** process spinning off a number of instances of the **Carry out a Tactical Review** process. Each of those instances will do its work and feed back to (/interact with) the **Carry out a Strategic Review** process.

This sort of thing is exactly what constantly goes on in organizations. During the development of a pharmaceutical drug, clinical trials are 'set going' to examine aspects of the safety and efficacy of the proposed drug, and they report back to the 'main process' with the results as and when they complete. Many clinical trials are in progress at any one time. That's the sort of concurrency that runs right through any organization. In Riva, we have precisely the language we need to describe what happens in the real world. We will say that 'process *A* activates process *B*', or that that 'process *A* instantiates process *B*'. They have the same meaning: 'An instance of process type *A* instantiates process type *B*'.

We must now say precisely what we mean by saying that one process instantiates another. We shall start by showing how it can be represented in a very minimal way.

Suppose that when we model **Carry out a Strategic Review**, we want to show **Carry out a Tactical Review** being *activated*. Figure 4.5 shows how we do this. In the activating process – **Carry out a Strategic Review** – we show an appropriate role (here *Strategic Review Board*) instantiating what we call the *lead role* of the activated process (here *Task Force*). The *Task Force* role is an 'abstract' role: it doesn't have pre-existing instances, in particular, such things do not appear on the organization chart; instances

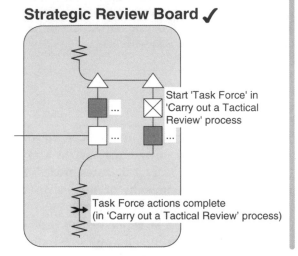

Part of the
Carry out a Strategic Review
process

Strategic Review Board ✓

Start 'Task Force' in 'Carry out a Tactical Review' process

Task Force actions complete
(in 'Carry out a Tactical Review' process)

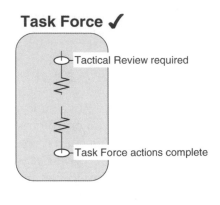

Part of the
Carry out a Tactical Review
process

Task Force ✓

Tactical Review required

Task Force actions complete

FIGURE 4.5 *Minimal process activation and interaction*

are only created as and when required and once they have done their work they disappear. The lead role represents the responsibility for the process: hence the term *lead role*. This reflects real life: when we decide we want to have a *Task Force* do something for us, we *create the responsibility* for that something: that responsibility is the lead role. Moreover, we can associate the instance of *Task Force* with the instance of the **Carry out a Tactical Review** process. So when we want to instantiate a process we simply instantiate its lead role – everything then follows.

When we then draw up the RAD for **Carry out a Tactical Review**, we naturally show the lead role, *Task Force*, and, to indicate that it has a preexisting instance (generated in **Carry out a Strategic Review**), we tick it with a ✓.

Note that we do not need to show what makes things start in the RAD for **Carry out a Tactical Review**: the ✓ is enough to indicate that the *Task Force* is already instantiated, so we can simply show what it does when it gets going. That said, we might choose to label the state at the start of the role with a suitable caption (e.g. *Tactical Review required*), simply to make this separate RAD that bit more free-standing so that it can be read on its own.

The RAD for **Carry out a Tactical Review** shows that the role *Task Force* carries out some activity and then, at some point, is in the state *Task Force actions complete*. This state is probably important to the Strategic Review process; the *Strategic Review Board* is not interested in how it is reached, only when it is reached, since it needs to pick up and respond in some way to the results of the *Task Force's* work. Figure 4.5 shows how we can represent that in a very minimal fashion.

Note how we use the ➤➤ to spot the completion of the *Task Force's* work in the activated process and how (in this model) we have used it to import the results of their work into the activating process. In other words, the activating process synchronizes with that state by waiting on a corresponding event (*Task Force actions complete*). We can think of this as a very bare way of representing a process interaction.

Note how the two resulting RADs can either be read entirely separately or be seen as a coherent pair, using this minimal amount of modelling to capture the activation and the subsequent interaction.

Suppose we didn't want to be quite so minimal in our model. For instance, on the model for **Carry out a Strategic Review**, we might want to explicitly show the final interaction with *Task Force* where results are returned to the *Strategic Review Board*. In particular, we would like to use the technique we developed above for showing the interaction between two processes, which is what this return of results is. The result would be something like Figure 4.6 in which the shared role *Task Force* appears on both RADs. In Figure 4.5 we used a ➤➤ to spot completion of the *Task Force* actions over in **Carry out a Tactical Review** and to import the results.

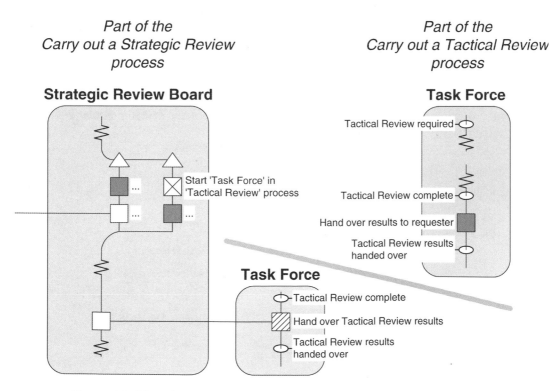

FIGURE 4.6 *More-than-minimal process activation and subsequent interaction*

In Figure 4.6 we use an explicit interaction to hand over the Tactical Review's results, and use our normal process interaction notation to tie this back to **Carry out a Tactical Review.**

Let's go one step further and suppose we are interested in the fact that the *Task Force* receives terms of reference when it starts. We might draw the RAD in Figure 4.7. Now we have still shown the role *Task Force* in the

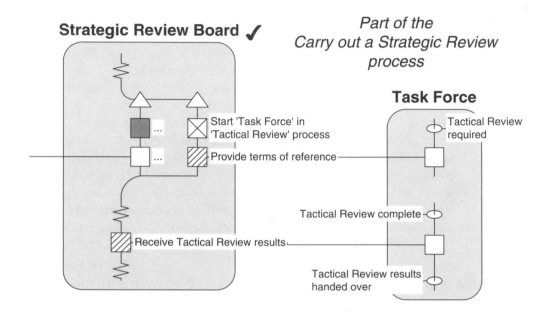

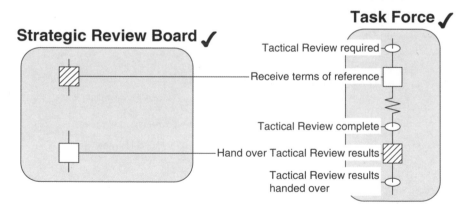

Part of the Carry out a Tactical Review process

FIGURE 4.7 *Full process activation and interaction*

Carry out a Strategic Review RAD, but we have added the interaction to pass the terms of reference to the *Task Force*, and have also moved the start boundary of the **Carry out a Tactical Review** RAD by adding that initial interaction for receiving terms of reference. Take a moment to check how the events and states have been used to tie the two RADs together whilst keeping them free-standing, and to observe the complementary nature of the way the relationships between the two processes are represented in the two RADs. Note the symmetry of the two RADs in Figure 4.7 and compare it with the minimalist model in Figure 4.5.

The minimal modelling of process activation

To model process activation in a minimal way, we proceed as follows:

- In the activating process A, instantiate the lead role in the activated process B.
- In the activated process B, identify (for information) the start state of the lead role and show that role as having a pre-existing instance.

Later, next to the water cooler

Pupil: I'm rather surprised by all this. What you're saying is that process A activates process B simply by instantiating B's lead role. That's all that is necessary for things to begin?

Tutor: Yes, it is. Like all things Riva, we try to keep things minimal. Let's think through what's happening. When I activate a process it's for a reason: I want something to happen. To make something happen, I create a responsibility for making it happen. The lead role is that responsibility. By instantiating the lead role we have created the responsibility. The process begins. We don't instantiate other roles or actions or whatever ... not until their time comes.

Having instantiated the lead role, the activating process A might immediately assign actor(s) and hand them resources – their props – in an interaction. We might choose to model this ... or we might not. By instantiating the lead role, we are – in effect – instantiating the activated process B. The instances of A and B then have their own lives, operating concurrently, interacting as necessary. In some situations the activated process instance will have a shorter lifetime than the activating process instance, perhaps 'living' only as long as is necessary to do a job and give the results back to the activating process instance. In Chapter 6, when we construct the organization's process architecture, we shall see how this sort of 'service' relationship underpins much of the dynamics in an organization.

Earlier in this chapter I said that organizational activity is the *operation of a changing network of concurrent, interacting process instances*. It should be clear now how such a dynamic network can be seen as the operation of process activation and process interaction.

ENCAPSULATION

Given that our functioning organization is a network of interacting process instances, we can see that hierarchical decomposition would be a totally inappropriate way of modelling organizational dynamics. To say that 'process A is subprocess A1 plus subprocess A2 plus ... ' is to ignore reality. It is hard to give any meaning to such a statement. What exactly do we mean by 'plus'?

However, when we draw a RAD there are occasions when we might wish to 'summarize' a whole mass of activity in a single black-box action, not wishing – in this particular perspective – to get into the detail of how it is done. Similarly, we might wish to summarize a complex interaction as a simple interaction, not wishing – in this particular perspective – to worry about the detail. We should therefore allow ourselves the possibility of such a modelling convenience. But note that this is a *modelling convenience*, and we are not pretending that processes are hierarchically structured. We must take great care not to imagine that we can accurately model a process by decomposition into smaller and smaller process 'units' or 'subprocesses'.

Opening up a black-box action

The question is 'What happens if we "open" a black box on a RAD?' Rather than talking about the 'decomposition' of a black-box action, I prefer to say that we are looking through a window and seeing part of the world from another, more detailed, perspective. We need to understand how that new perspective relates to the black box on the original RAD. In particular, we must remember that an action (like an interaction) takes place over some period of time and at a particular place in the process, and so we need to understand especially the temporal relationships between the two perspectives, the 'main' RAD where we see a black box and the 'encapsulated' RAD which we see when we look through the window.

For instance take the action *Produce design* back in Figure 2.2. Suppose we open this black box. When we look through it, we might find a whole world of activity involving new roles such as *Client, User, Quality Assurance*, and *Chief Engineer*. These do not appear on the first RAD, nor are they 'part of' the role *Designing*. But they are all 'encapsulated' in the action *Produce design*. In treating *Produce design* as a process in its own right, we are starting to look at new parts of the world, parts that we were not interested in when we were drawing up the model in Figure 2.2. So we are now going to draw two free-standing RADs whilst showing the relationship between them.

To understand how we open up a black box on a RAD, let us look at a simple example. Suppose we are a life insurance company and suppose that the way the Underwriting Manager handles an application for an insurance policy depends on the office where the application is received; each office has its own procedure for getting partner approval for something. If we were not concerned with the detail of each office's procedure but did want to identify which procedure they used then we might draw the relevant part of **Handle an Insurance Claim** as in the upper part of Figure 4.8, showing the *Underwriting Manager* using procedure P12 in the London office, procedure P13a at the New York office, and so on, with each shown as a black-box action.

Suppose now that we want to go into detail about the different office procedures but, for modelling convenience, to show it on its own RAD. How would we do this? Basically, we draw a new RAD for each office procedure's black box. In other words, we treat each black box as if it were a complete process and give it a RAD of its own; for instance, action P12 on the main RAD becomes **Process P12** on its own RAD. Remembering my earlier warning about the dangers of decomposition, we have to ask what precisely we mean by saying 'Action P12 has its own RAD as **Process P12**.'

Firstly, what 'starts' **Process P12**? In **Process P12** we will naturally expect to see the role *Underwriting Manager* from the 'main' process, the role that carries out action P12 and, moreover, we can assume that one instance of *Underwriting Manager* exists when process P12 runs. This is the 'lead role' of the process: it is the one that pre-exists and that picks up the thread at the beginning. So the activating condition of action P12 is also the state at the start of the lead role in **Process P12**. We will show this by labelling the states accordingly: *Partner approval required.*

Secondly, when we know that P12 is complete, the *Underwriting Manager* is in the state *Partner approval obtained*. This too must appear on the RAD for **Process P12** and we have shown this in the lower part of Figure 4.8. But when we say that 'Action P12 has its own RAD as **Process P12**,' do we require that *all* of **Process P12** has to be completed before the black box for action P12 is deemed complete and the main process can proceed? If the answer were 'yes' it would, to use software jargon, be like treating process P12 as a 'subroutine' which must 'complete' before 'control passes back to' the main process. Again, there is a temptation (especially for the software engineer) to impose a tidy block-structured simplicity on the world; but the world is rarely so clean. Instead we are saying that **Process P12** must reach *a specified state* before action P12 is complete. We recognize that there will be some state reached during **Process P12** which is the same as the post-condition of action P12.

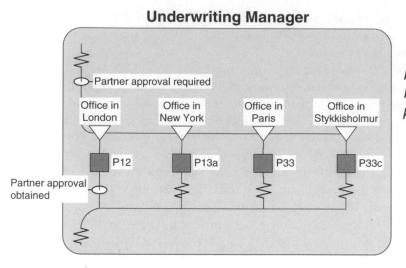

Underwriting Manager

Part of the Handle an Insurance Claim process

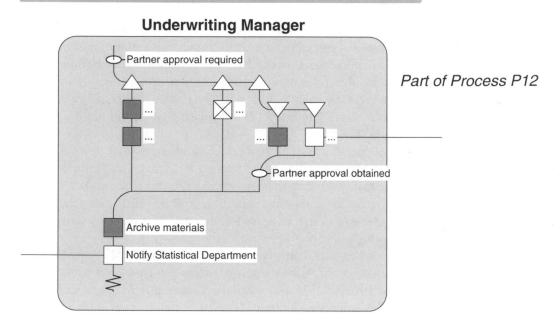

Underwriting Manager

Part of Process P12

FIGURE 4.8 *Action P12 and Process P12*

In the second RAD fragment in Figure 4.8 we show part of the RAD for **Process P12**: the process might be quite complicated but at some point that same state – *Partner approval obtained* – is reached, at which the thread in **Handle an Insurance Claim** can proceed, even though there is clearly further procedural activity in **Process P12** before it finishes: the Statistical Department must be informed, archiving must be done, and so on. The *Underwriting Manager* in **Handle an Insurance Claim** does not have to wait for all these things to be done before proceeding. (Process elements have not been labelled in Figure 4.8 where they are not pertinent to the example.)

So there is no sense in which **Process P12** and action P12 are the same thing. All we can say is that some of **Process P12** constitutes action P12.

Later, next to the water cooler

Tutor:	**To demonstrate why I am nervous about this idea of encapsulation and the danger of imagining there are things called 'subprocesses', let me tell you my suspicion about this Handle an Insurance Claim process: it is that the partners in this insurance company have a stream of requests for approval coming to them, and that that stream of work has to be managed, prioritized and perhaps resourced, with approval requests being forwarded to partners on the basis of availability, loading etc. In other words, that stream of requests is managed somewhere.**
Pupil:	**That sounds like a process itself to me.**
Tutor:	**Exactly. So the RAD we have drawn is a lie: it shows none of that flow management. If we are in a process improvement project and we don't show it, we can never recognize it as a potential bottleneck. If we are designing a new process and don't show it, we will have a gap in our processes. If we are planning to enact these processes on some form of BPMS, then we shall need to be very clear where such case management occurs. So by thinking in terms of 'subprocesses' we have made a major modelling error. That's one of the reasons why I am nervous about encapsulation and its use.**
	Riva's solution to this is the *process architecture*. This is the way we find all the processes in the organization, in particular the flow management processes. In Chapter 6 we shall see how to expose all the processes and, as a result, we should see that encapsulation should only be necessary – as a modelling convenience – in a very few cases.

Modelling action encapsulation

Let's summarize the steps in representing the encapsulation of an action:

- In the main process RAD, show the black-box action A to be opened in the appropriate role R, label its pre-state, say ST, and label its post-state, say SP.

- In the encapsulated process RAD, label the state at the top of the main thread of R ST. Show the state SP at the appropriate place in R.

Spend a moment thinking through how the two states – both shown on both RADs – tie the two processes together, much as we might solder ends of wire together to connect electrical circuits.

Finally, let's note that there is a special symbol for an action that has an encapsulated process for it on another RAD:

 Get partner approval

When we move from the main process to the encapsulated process we might find it useful to work with a sub-role of the role in the main process. So, although we might have the role *Accounts* carrying out the action *Review Budget* in the **Prepare the Annual Budget** process, when we model the encapsulated 'process' **Review Budget** in its own RAD, we might show the *Chief Accountant* starting it off – a role that is 'within' *Accounts*. Since we want the RADs to be free-standing, we would tie them together with some extra captioning on the action in the main process RAD, as in Figure 4.9. The dotted lines emphasize the way in which states are used like solder points to connect the main and the encapsulated processes.

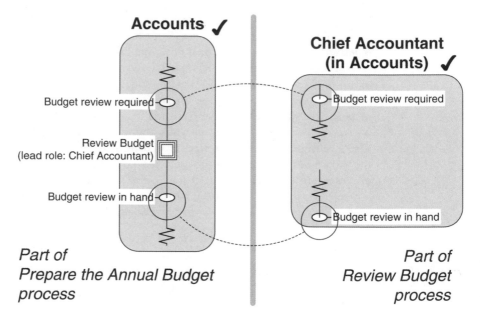

FIGURE 4.9 *Using a sub-role in an encapsulated process*

Opening up an interaction

In a RAD, we might choose to summarize a complicated interaction in a single element as a modelling convenience. As an example, take a simplification of the interaction I used earlier: 'Two parties meet to discuss, negotiate and agree the price of a piece of work, drawing up the agreement as a legal document and obtaining financial securities from a bank.' On a RAD, we might choose to represent this as an atomic interaction between the roles *Buyer* and *Seller*, atomic in that we are not interested – in that model – in any further detail; we simply want to say that *Buyer* and *Seller* have that interaction with that result and we don't mind how they do it.

Suppose we now choose to 'open up' this interaction. Rather than showing just more detail of what happens between *Buyer* and *Seller* (in the way that we noted when we looked at interactions as 'conversations for action' in Chapter 2), we might look further and find other roles involved, roles which did not appear on the first RAD: *Bank Manager*, *Lawyer*, and *Auditor* for example; we might also find new actions that they carry out and new interactions between them. We have not 'decomposed' the interaction: we have opened it up like a window again and looked at this part of the world from a new angle, an angle which introduces new roles and activities, all of which were of no interest to the first RAD.

As when we opened black-box actions, we need to understand the relationship between the interaction and its 'expansion'. Not surprisingly, we do it by equating the pre-states of the two part-interactions with corresponding starting states in the roles in the expanded process. The post-states of the part-interactions are similarly dealt with. Spend a moment tracing through the example in Figure 4.10. In the expanded **Finalize a Sale** process, much can happen between the initial and final states.

Part of the Buy a House process

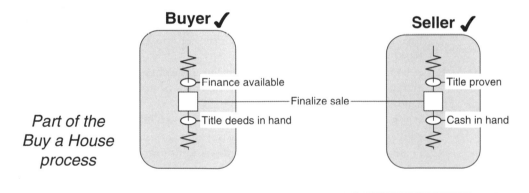

Part of the Finalize a Sale process

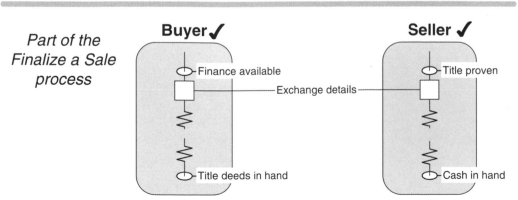

FIGURE 4.10 *Encapsulating an interaction*

Modelling interaction encapsulation

Let's summarize the steps in representing the encapsulation of an interaction between roles R1 and R2:

1. In the main process RAD:

    (a) Show the part-interactions P1 and P2 of the interaction to be opened in R1 and R2 respectively.

    (b) Label the pre-states of each part-interaction, say preP1 and preP2.

    (c) Label the post-states of each part-interaction, say postP1 and postP2.

2. In the encapsulated process RAD:

    (a) Label the state at the top of the main thread of R1 preP1.

    (b) Label the state at the top of the main thread of R2 preP2.

    (c) Show the states postP1 and postP2 at the appropriate places.

Again, spend a moment thinking through how the four states – all shown on both RADs – solder the two models together.

KEY POINTS

Don't use encapsulation.

If you think you must, don't until you have fully understood the process architecture and cannot find the encapsulated 'process' in it.

If you still think you must, ask first whether you have uncovered another UOW as described in Chapter 6.

If you finally do, use states to show how the beginning and end of the action/interaction being encapsulated translate into the 'expanded' process.

When you have done it, remember that encapsulation is only a modelling convenience and probably doesn't reflect anything in the real world.

Finally, reconsider whether you really should have done it!

---

5 The three basic process types

Describes the three main types of process – the case, case management, and case strategy processes.

INTRODUCTION

We now have in place the concepts and vocabulary for describing individual processes, and for modelling the two sorts of relationship between processes: activation and interaction. In the Introduction we – rather vaguely – observed that there are three types of organizational activity: what I called coal-face, management, and strategic activity. In this chapter, we shall look at three types of process and give them more precise names and much more precise meanings:

- case processes;
- case management processes;
- case strategy processes.

Our hypothesis will be that everything that happens in the building is part of a case process, or a case management process, or a case strategy process. We shall use this hypothesis in the next chapter when we look at how to chunk all the activity in the building: how to draw up the process architecture of the organization.

THE CASE PROCESS

Units of work and cases

Suppose we are looking at the department in a life insurance company that deals with new business and, in particular, applications for new insurance policies; and suppose we are particularly interested in what happens to an application for a new life insurance policy, from the point at which it is received by the company to the point where some outcome is reached with the prospective customer. We can think of the customer application as the *unit of work* (UOW) for this department: it is the unit in which work arrives and is dealt with, and every application is dealt with in the same, standard way.

If we walked into a software house, we would find them taking on pieces of work for a client – which are typically called projects. The 'project' is the UOW of the software house. Each piece of work is one *case* of the UOW 'project' that typically starts with the award of the contract and finishes with acceptance of the software by the client. At any one time the software house will have many projects (cases) in progress, all at different stages of some standard project lifecycle.

In a pharmaceutical R and D company, each potential new drug compound has a development lifetime that takes it from the point where it is determined to have some possible therapeutic effects, to the point at which it obtains approval from the regulatory authorities to be put on sale or is dropped. Here the UOW is the 'compound'. During the development of the drug compound, the company carries out clinical trials to determine its efficacy and safety. The 'clinical trial' is also a UOW, in fact a major work item dealt with by the Clinical Department.

Let's think of some more UOWs and the groups for whom they are UOWs:

- a house purchase, for a solicitor's office;
- a purchase order, for a supplies company;
- a marketing campaign, for the Marketing Department;
- a production batch, for a factory;
- a customer complaint, for the Customer Services section;
- a product line, for a manufacturer;
- an operation, for a hospital;
- a patient, for a hospital;
- a student, for a college;
- a course module, for the Physics faculty;
- an exam paper, for a university course;
- a meeting, for the committee secretary;
- a blocked drain, for the Maintenance Department;
- a clinical trial, for a pharmaceutical company;
- a phone call, for a call centre;
- a blood donation, for a blood bank;
- a stock purchase, for a broker;
- a donation, for a charity;
- an amendment to a purchase order, for a supplies company;
- a company, for a conglomerate;
- a request for new staff, for the HR department;
- the annual budget, for the Board;
- a generator, for an electricity supply company;
- a lease, for a lease owner;
- a customer, for a services company;
- a building, for a construction company;

- a building, for a property company;

- a project report, for a project manager.

Take a moment to jot down all the UOWs you can think of in your part of the organization, or even in your home. A child, for the parents? A meal, for the cook? A visit to the supermarket, for the shopper? A journey to the office, for the commuter?

The list above shows how varied UOWs can be:

- Some are solid, physical things with a lifetime that we can easily see: a generator, a building, a patient, a customer. During their lifetime we shall 'look after' them.

- Some are less tangible: a purchase order, a customer complaint, a course module. Their lifetime is a little harder to define but during that lifetime we shall 'deal with' them. (They might have a paper or other physical manifestation, but that is incidental.)

- Some are rather abstract: a project, a clinical trial, a meeting. They have a duration rather than a lifetime and begin and end when we say so. We might say that we 'do' them.

- Some are very abstract: an amendment to a purchase order, a product line.

UOWs and case processes

This sort of situation is common; perhaps we might claim it is the way all organizations work. Work comes in 'cases' or 'episodes', each needing to be dealt with in a standard way (assuming we don't behave randomly when each case arrives). The moment we say 'dealt with in a standard way', we have recognized that each case follows the same process. We shall call that process the *case process*. This will be the process that 'looks after', or 'deals with', or 'handles' a case during its lifetime. When a case 'arrives', 'comes into our area' or 'lands on our desk', however we put it, we start the case process working on it. In fact, given the concepts and vocabulary we developed in Chapter 4, we can now say that we 'activate the case process' for each new case. We might even say that we 'instantiate the case process' for each new case.

So at any one moment, we could look around and see perhaps many instances of the case process in progress: one for each case currently being dealt with. The office supplies company has 1,222 purchase orders 'in hand'. The call centre is currently 'dealing with' 34 calls. The Board is 'working on' the annual budget. The pharmaceutical drug company has 15 compounds 'in the pipeline'. The Clinical Department is 'running' 87 clinical trials. In everyday speech we have many ways of saying 'A case process instance is operating.'

Each case will be at a different stage in its case process: for example, in our pharmaceutical research company, compound A might be in 'first in man' trials with human volunteers, compound B might be undergoing major Phase III trials with thousands of clinical patients, and compound C might be awaiting regulatory approval at the end of its development process.

In some situations, cases might come just one at a time: there is only one annual budget each year and we finish work on one before we start on the next.

Naming case processes

Because the three process patterns will prove to be central to our thinking about processes, it will be useful to be able to distinguish them by the names we give them. This might feel over the top at first, but as we use the convention more and more, we will find that it helps keep our minds focused on the case process as opposed to the case management process, or on the case strategy process as opposed to the case management process and so on.

To emphasize that we are talking about a case process, we shall always start by naming a case process with the words **Handle a** or **Prepare a**. We use **Handle a** where the UOW could be seen as the *input* to the case process or the thing that in some way triggers it:

- **Handle a purchase order**.
- **Handle a customer complaint**.

We use **Prepare a** where the UOW could be seen as the *output* or the outcome in some way. So we would have:

- **Prepare a production batch**.
- **Prepare a project report**.

Where the UOW is really neither an input nor an output, use **Handle a**.

- **Handle a marketing campaign**.
- **Handle a house purchase**.

Already, these names are emphasizing that the process is about *one* thing: *one* purchase order, *one* production batch, *one* marketing campaign. By all means choose a different name later in the analysis: **Run a marketing campaign**, or **Make a production batch**, or **Deal with a customer complaint**, or **Satisfy a customer order**. But try starting with the stilted names before moving to something else. If there is only one case of the UOW, we replace 'a' by '**the**' in the name.

Bounding case processes

So in Riva, we name case processes in a way that emphasizes the UOW. Other naming conventions that you might come across emphasize another important aspect, their end-to-end nature: *Order to cash, Engage to close, Transact to fulfil, Build to order, Plan to produce, Résumé to work, Goal to reward.*

Clearly, a case process starts when the case 'arrives'. A case 'lands on the desk' and the case process starts. We'll need to be thoughtful about precisely when that happens. In workshop situations – as we shall see later in Chapter 8 – I like to ask the question 'How do I know I've got a case?'

- At what moment do we know a house purchase has started? Is it when an offer to buy is accepted? Or when the offer is first made? Or when the potential buyer visits the property? Or when the house goes on the market?

- When does a customer complaint start? When the customer has finished filling in the complaint form? Or the moment they stand in front of the Customer Service desk and announce that they have a complaint? Or when they first joined the queue at the desk?

- When do I become a customer of the corner shop? Is it when I have made my first purchase? Is it when I walk in the door? Is it when I move into the locality?

We cannot answer these questions without understanding the larger context, the architecture within which this one process sits. So we must defer an answer until Chapter 6.

How long does a case last? How long is a piece of string? A phone call to the call centre could last twenty seconds; a pharmaceutical drug compound trial can last twenty years. So there is no minimum or maximum *lifetime* of a case.

Where does a case process 'finish'? My workshop question is 'How could I tell you've finished?' or 'What state are things in when the case has been dealt with?' We need to be just as thoughtful here as with the start point. When do I cease to be a customer of the corner shop? When I have paid for the goods? When I have left the shop? When I have left the locality? Again, we must know more about the larger context before we can answer this question.

It's very easy to assume that the end point of **Handle a customer complaint** is something like 'The customer has gone away happy.' That would certainly be a *desirable* outcome of the case process but it is by no means the only possible outcome. How about 'The customer drops the complaint' or 'The customer rejects our offer and takes the case to the industry ombudsman'? These are other possible outcomes, and we can imagine the handling of a customer complaint ending – as far as we and our model are concerned – with any one of these three possibilities.

By thinking through the possible outcomes, we are often forced to reconsider the name of the case process – and hence the way we think about it. Take a UOW such as 'expense claim'. It would be all too easy to think that the case process should be called **Approve an expense claim** ('Claim to payment'?) The absurdity of this is apparent when we realize that one possible outcome is that the expense claim is rejected. The process is not to *approve* an expense claim but to *handle* an expense claim. This is a mistake that is often made, and of course it can easily blinker our understanding of the process.

The traps can be quite subtle. **Satisfy customer order** ('Order to cash'?) is exactly the customer-oriented and success-oriented name we might like to give to a process. But some customer orders cannot be satisfied and have to be rejected. **Handle a customer order** leaves open the fact that our case process must deal with ill-formed orders, and orders from customers we choose not to do business with, and orders sent in error. By using the neutral phrases **Handle a ...** and **Prepare a ...** we leave the other possibilities open and we don't blinker our thinking.

> ### KEY POINTS
>
> A case process is the process that deals with one case or instance of a UOW.
> A case process should be named Handle a ... or Prepare a ... , depending on whether the UOW is an input/trigger or an output/outcome.
> A case process typically has a single starting point, corresponding to the 'arrival' of the case.
> A case process can have one or many possible outcomes.

Let's go back now and examine some of those UOWs we identified earlier on. Some of them might feel a tad strange when we put **Handle a** in front of them. For instance, what are we to make of the case process **Handle a customer**? To start to answer this we must start with the question 'What is the lifetime of a customer?' At what point does a new 'customer' case arise? If our organization takes a long view of customers, then a new case might start the moment we have a name and an address of someone who might buy something from us. Our **Handle a customer** process is now about getting that person into our store, giving them a satisfying retail experience, ensuring they return often, sending them special offers, rewarding frequent purchases etc. As far as my local supermarket is concerned, I exist as a customer – as a case – even when I am not in their store. My instance of their **Handle a customer** process will probably only end when I tear up my loyalty card and send them the pieces. On the other hand, I can think of other shops who might have a **Handle a customer transaction** process, but whose **Handle a customer** process is quite empty.

THE CASE MANAGEMENT PROCESS

The flow of cases

For each UOW there is a case process. And, at any one moment, there may be many cases of the UOW and hence many instances of that case process in progress. Given all this concurrent activity, possibly sharing resources or facilities, the organization will need to manage the flux, dealing with issues of planning, scheduling, resource management, task allocation, making go/no-go decisions, reporting, and so on. But when we take the case-oriented view, we are putting aside all of these concerns and concentrating simply on what happens to a single case.

For any UOW we can therefore expect to find two processes: one for the case and another for *case management*. We shall refer to the latter process as the *case management process*. The two processes will of course interact, but separating them is vital for effective process design and analysis.

Before we look more closely at how we represent the typical relationships between a case process and a case management process, let's take a closer look at the sorts of things that go on in case management processes.

The contents of a case management process

In a case management process we shall expect to see actions to do with:

- planning;
- reporting;
- monitoring;
- scheduling;
- resourcing;
- prioritizing;
- negotiating;
- reconciling.

So, we shall expect to see roles such as:

- Boards;
- managers;
- management teams;
- management committees;
- supervisors;
- progress chasers;
- planning teams;
- programme support offices;
- monitoring groups.

The case process is normally quite straightforward, being by definition the process which takes a single case from 'birth' to 'death'. As such, it will tend to have one trigger corresponding to the birth of the case, and one or more alternative outcomes corresponding to its different forms of death. For instance, the **Handle an insurance application** case process might be triggered by the arrival of an application and have two alternative outcomes: application accepted and application rejected.

Case management processes are never so simple and rarely single-threaded. The nature of management is that it responds to many different situations and intervenes as necessary, and is proactive in many other situations. We can therefore expect the full-blown case management process to have many triggers corresponding to the different stimuli, each with its own outcome(s). We must not imagine that every case management process will contain all of the following components – indeed some case management processes are trivial or even null – but here is a list of typical components of a case management process:

- Dealing with a request for a new case, i.e. for a case process instance to be started.
- Negotiating with a requester if the request for a new case cannot be met (at the required resource cost or timescale).
- Monitoring the progress of current case process instances.
- Hearing about and dealing with the completion of a case process instance.
- Hearing about and dealing with exceptions and failures from case process instances.
- Determining what resources should be assigned to the acting of which case process instances.
- Adjusting the resources currently allocated to existing case process instances as loading changes.
- Dealing with requests for shared resources from a case process instance, e.g. actors for new role instances.
- Dealing with requests, typically from other case management processes, to negotiate about priorities on services being supplied by those case management processes ('escalation').
- Dealing with instructions from the case strategy process about how case management is to be done (see below).
- Receiving budgets or resources from 'superior' management processes for providing the service (assuming this is a UOW that is provided as a service, e.g. an invoice).
- Recording and analysing trend data, and responding to the results.
- Assessing immediate resource trends and estimating near-term resource requirements.
- Assessing exception trends and redefining the case process for process improvement.
- Auditing the behaviour of case process instances.

The case management process is essentially taking responsibility for the flow of case process instances. This is key. When a new case of a UOW comes along and needs an instance of the case process to be started, *it is the case management process that must be asked to start it*: it is the process that decides when – amongst all the other instances contending for time and resources – this new instance is to start. It is the case management process that monitors all the active case process instances, and manages resourcing and scheduling amongst them.

A case management process might batch cases until there are enough to start work on them. It might start them in strict order of arrival. It might juggle their ordering depending on their relative priorities. It might move resources from one to another. It will sort out conflicts over priorities between competing cases. We might summarize all this by saying that it 'manages the flow of cases'.

Naming case management processes

In the same way that we chose 'neutral' names for case processes with the **Handle a** and **Prepare a** prefixes, we shall use a similar rule for case management processes: we shall start them with the words **Manage the flow of**. For example, we might have the following processes:

- **Manage the flow of purchase orders**.
- **Manage the flow of customer complaints**.
- **Manage the flow of production batches**.
- **Manage the flow of project report**s.
- **Manage the flow of marketing campaigns**.
- **Manage the flow of house purchases**.

Once again, the purpose of this convention is to concentrate our minds, in particular to help the separation of concerns: dealing with the individual case is the responsibility of the case process; managing across the cases is the responsibility of the case management process.

Whilst we should start with this stilted name, we shall feel free to choose a more meaningful name later, if it's appropriate.

> ### KEY POINTS
> In principle, every UOW has a case management process that manages the flow of cases of that UOW.
> The case management process exists in only one instance.
> It contains management roles carrying out management-related actions to do with scheduling, prioritizing, and resourcing.
> Requests for new cases are always directed at the case management process.
> It activates case process instances when required.
> It interacts with its case process instances when required.

THE RELATIONSHIP BETWEEN CASE PROCESS AND CASE MANAGEMENT PROCESS

We have seen that each UOW has a case process and a case management process, and we have a good sense of the separation of concerns. A proven strength of this approach is that it gives equal weight to how we deal with day-to-day work (cases) and how we manage that work (case management). Let's look at how the relationship works in practice.

The service relationship

Suppose we are in a factory and we have a UOW called the 'order'. Down on the production line the UOW is the 'production batch'. An order arrives. An instance of the **Handle an order** process runs. (How did that happen?) At some point it will determine that a batch must be produced to satisfy the order. Of course the factory already has a number of existing batches going through production. Our instance of **Handle an order** cannot simply start up an instance of the **Prepare a batch** process – the new batch needs to be worked into the production schedule. This is precisely the responsibility of the case management process, **Manage the flow of batches**. So our instance of **Handle an order** must ask **Manage the flow of batches** to add the new batch to the schedule, and, when the time is right, it is **Manage the flow of batches** that will start (/instantiate) the **Prepare a batch** process. The **Prepare a batch** process instance then carries out the responsibility of making that one batch at the time and with the resources specified by **Manage the flow of batches**. Put simply, it will make the batch. We can think of **Handle an order** as the 'customer' of 'supplier' **Prepare a batch**, and when the batch has been made it will be supplied by **Prepare a batch** to **Handle an order**.

Let's restate this very precisely in our Riva vocabulary:

1. A **Handle an order** instance interacts with the **Manage the flow of batches** instance to request a batch.

2. **Manage the flow of batches** schedules the requested batch.

3. **Manage the flow of batches** activates **Prepare a batch** at the appropriate moment.

4. Once the batch to satisfy the order is ready, the **Prepare a batch** instance interacts with the requesting **Handle an order** instance in order to hand over the batch. That **Prepare a batch** instance can then finish.

Figure 5.1 shows these basic dynamic relationships between the three processes. (Let's introduce a couple of abbreviations: we shall occasionally use CP for case process and CMP for case management process.) A rounded rectangle denotes a process. We can tell from its name whether it is a case process or a case management process. An arrow with an I in a

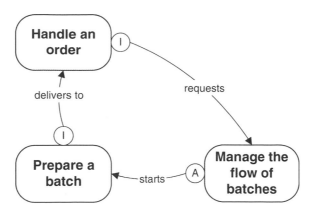

FIGURE 5.1 *The basic service relationship*

circle at one end represents an interaction between two processes; the arrow goes from the process that initiates the interaction. An arrow with an A in a circle at one end represents activation; the arrow goes from the process that does the activation to the one that is activated. Spend a moment checking through steps 1–4 above against Figure 5.1. At the 'close' of the relationship in step 4, the service case process 'delivers' to the requesting case process. We use the word 'delivers' in a very general sense: delivery might take the form of physical delivery – a transfer of goods from one role to another – or it might just be a notification that the service is complete.

Life is generally not as simple. Just asking for a batch of 20 units to be ready on 15 June does not guarantee getting a batch of 20 units on 15 June. Is there the capacity given all the other batches in the queue? If not, what has to happen? In reality, there will be a negotiation between the 'customer' **Handle an order** and **Manage the flow of batches**, which might involve some reworking of the schedules to achieve the target. And, of course, other batches may be affected, requiring wider negotiation. How is that negotiation done? Typically, **Manage the flow of batches** must negotiate with **Manage the flow of orders**. Ultimately, only the CMP for orders can resolve inter-order issues. Let's add negotiation to the basic service relationship: Figure 5.2.

We can also add some other basic management relationships between CP and CMP: the CMP will monitor the CP through a process interaction; in some situations the CMP may intervene in the CP, perhaps to notify it of resourcing or scheduling changes, perhaps in extreme situations to stop it; the CP will report status and exceptions and management-related problems to the CMP, and so on. This gives us the general service relationship shown in Figure 5.3.

Of course, the full set of relationships shown between **Manage the flow of batches** and **Prepare a batch** also exists – in the general case – between

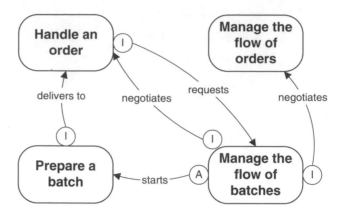

FIGURE 5.2 *The basic service relationship with negotiation*

Handle an order and **Manage the flow of orders**. (Yes, that original instance of **Handle an order** was created by **Manage the flow of orders**!)

There is another possible complication: the service might be used by the cases of more than one UOW. Suppose that the R and D Department is working on potential new products. We can think of a 'potential product' as a UOW. It therefore has a case process, **Handle a potential product**, which carries out the R and D necessary to bring the new product to production capability, and a case management process, **Manage the flow of potential products**, which ensures that the R and D pipeline is working optimally. Imagine that, as part of product development, **Handle a potential product** requires test batches to be made in the factory on the same production line. To get those batches made, it too will have to knock on the door of **Manage the flow of batches** and ask to have its batches scheduled in. As well as reconciling the demands for batches to satisfy orders, the CMP must now add to the mix requests for batches for R and D.

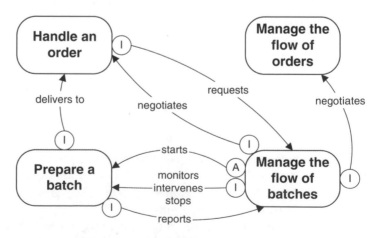

FIGURE 5.3 *The general service relationship*

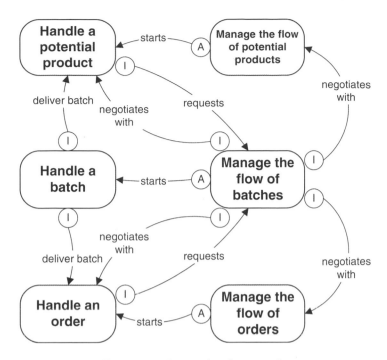

FIGURE 5.4 *Contention for a service*

In Figure 5.4 we can see the resulting process structure: note how **Manage the flow of batches** now has requests for batches coming from two places, and how in some cases it may need to negotiate with the corresponding CMP to reconcile problems. We can easily imagine some of the tensions that might arise between these, let alone the question of which customer's orders are more important than others. The CMP **Manage the flow of batches** becomes the focus of these issues, the place where they are sorted out. In the general case, any CP wanting a service must knock at the door of the appropriate CMP to obtain it.

This illustrates even more clearly the importance of the way in which Riva lets us cleanly separate issues to do with the single case from issues to do with the *set* of cases in hand at any one time. When we are designing or modelling the case processes, we know that we can concentrate on how a single batch is made, or how a single customer call is answered, leaving matters of scheduling, prioritization and resourcing where they belong: in the case management process.

Finally, let's take another example. During the development of a new drug compound, clinical trials are carried out. Many pharmaceutical companies outsource clinical trials to external Clinical Research Organizations (CROs). CROs provide a 'clinical trials service'. In our terminology, **Develop a compound** interacts with the **Manage the flow of clinical trials** process at the CRO when it needs a new clinical trial. That process has responsibility at the CRO for scheduling the clinical trial and, at an appropriate moment, starting it, i.e. activating its **Do a clinical trial** case process. Once the clinical trial is complete, the results are returned by the CRO's **Do a clinical trial** to the pharmaceutical company's **Develop a compound** and the service is complete.

The task force relationship

Let's stand back from what we have just done on the service relationship. The process for satisfying an order needs a batch to be made. It goes to the case management process for batches to request that service. Subsequently the service is delivered. Key to this is the fact that making batches is a *permanent service offered by someone else*. Anyone wanting a batch uses this service. The service operates independently of the processes that use it.

But when we need a job done we might not always go to a separate service whose job it is to provide it: we might set up the means to do the job ourselves. Let's call those means a task force. It's rather as if we ourselves have the case management process for the service.

Here's an example. Imagine a software house that carries out system development projects for clients. They no doubt have a case process called **Run a project**. Each project is planned and carried out as a number of separate work packages for specifying the system, designing parts of the system, developing the many software components, testing those components, and so on. The work package is (by definition!) a UOW so the software house will have a **Do a work package** case process too. But when a project wants a work package doing, it doesn't go to some independent 'work package service' to get it done; in particular, it doesn't interact with a separate **Manage the flow of work packages** process that deals with work packages from all projects. Managing the flow of work packages is something that the project *does for itself*: it's something that goes on inside – is indeed part of – **Run a project**. Each project itself manages the flow of its own work packages. Each work package is done by a little task force set up to do it that follows the standard process for work packages. There is clearly still case management to be done for work packages: they have to be scheduled, prioritized, resourced and so on, in the usual way. But that last sentence simply describes project planning and control ... which are part of **Run a project**. The underlying relationship between the processes is the same as for the service situation, but with one difference: in the service situation, **Handle an order** and **Manage the flow of batches** were independent processes, which interacted when an order required a batch; in the task force situation, **Manage the flow of work packages** is part of – is *encapsulated* in – **Run a project**.

Our general relationship diagram for the task force relationship looks like Figure 5.5. The only change from the service relationship diagram in Figure 5.3 is the 'E' for 'encapsulation' instead of the 'I' for 'interaction' where **Run a project** requests a work package from **Manage the flow of**

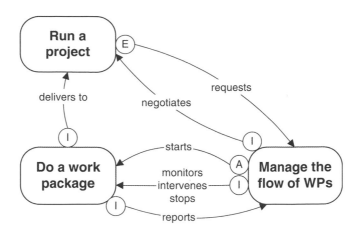

FIGURE 5.5 *The general task force relationship*

work packages. Note that there is no negotiation with what would be **Manage the flow of projects**: the management of the flow of work packages is 'inside' the running of the individual project that wants it, so that sort of negotiation across projects would never occur.

This task force relationship crops up in a number of similar situations. Take the concept of a 'programme of projects'. A business might establish a programme (a UOW, of course) to make some major business change. Carrying out that change will mean undertaking a number of separate, concurrent, and coordinated projects (another UOW) to, say, reorganize the department concerned, move staff and offices to a new location, recruit new staff skilled in some new area, update the information systems, and handle HR issues for existing staff. The programme will manage the set of projects itself; it will not go to some separate service that does projects. Put in our terms, we would say that **Run a programme** encapsulates **Manage the flow of projects**.

We saw how a large pharmaceutical company will have a portfolio of potential drug compounds in development – its 'pipeline'. 'Portfolio' and 'compound' are UOWs. In the process **Run the portfolio** new compounds will be identified and added to the portfolio, some will be removed from the portfolio if they do not have the necessary promise, and priorities will be changed between competing compounds in the portfolio. All of these are of course aspects of **Manage the flow of compounds**, and we can see the same task force pattern as in the programme of projects and the project of work packages. No external service is used for the UOW required: case management is done 'in-house'.

Let's return to our pharmaceutical company which outsources its clinical trials to a CRO. Fifteen years ago it might have done its own clinical trials. Instead of having a service relationship to get its trials done, it would have had a task force relationship and done its own case management.

> **KEY POINTS**
>
> When one case process wants a case provided by another case process, it must ask the corresponding case management process.
>
> The case management process is responsible for scheduling, prioritizing and resourcing cases, and for activating the case process when appropriate.
>
> Where necessary, the case management process will negotiate amongst contending case processes via their respective case management processes.
>
> The case management process can monitor, intervene in, and even stop a case process.
>
> Where a UOW is supplied as a service, its case management process operates independently.
>
> Where a UOW is supplied via a task force, its case management process is encapsulated in the requesting case process.

Later, next to the water cooler

Pupil: As I understand it, the name 'task force' is used because a wholly new group might be created to carry out the case process instance. But couldn't this occur in the 'service function' as well? When the service function gets the request it will instantiate the case process in exactly the same way. Both styles seem to have features of task forces.

Tutor: Yes, the two situations might be implemented in the same way but the thing that differentiates them is that, in the task force situation, the case management is effectively done by the requesting case process: I don't expect a service provider to do it. Imagine the situation where you have a project (an instance of the Run a Project case process). To get your project done, you break it up into work packages. Each of these is done by the Do a Work Package case process. Think it through.

Pupil: Well, it's clearly a task force situation as I do the management of the work packages myself, as part of my Project case process. There's certainly a flow of work packages – many are going on at one time. So there is case management to be done and there is a case management process.

Tutor: Exactly. But the case management of work packages isn't done by a service provider – it really is part of your case process for the project: you manage your own work packages within the project. The case management process for work packages is 'within' your case process for the project. It would be different if you subcontracted work packages to a service provider. In that case they would do case management for you, and your work packages would have to take their turn amongst a lot of others from other users of the service.

Pupil: This makes me think that whether a UOW is dealt with as a service or as a task force is a matter of design. I could choose to deal with it myself or I could ... outsource it.

Tutor: Right. We could see this as a re-engineering opportunity: an inflexible centralized service could be disbanded in favour of giving individual groups the freedom to set up their own task forces as and when they require them; or on the other hand, inefficiently replicated DIY processes could be replaced by a central service achieving economies of scale. The world is full of people swinging from one to the other. Most recently, outsourcing of non-core processes has been all the rage: in Riva terms, people have been giving the case processes and case management processes for what we shall call *designed* UOWs to others to do for them, whilst keeping the case processes and case management processes of (what we shall call) their essential UOWs in house. That's the Riva definition of outsourcing.

MODELLING THE CP–CMP RELATIONSHIP AT RAD LEVEL

We now know, at the process level, what form the relationship between CPs and CMPs takes: we have the generic pictures in Figures 5.3 and 5.5. What do these relationships look like in the RADs concerned? Again, generically, we know that we shall be using activation when the CMP wants to start a CP, and an interaction when a CP wants to deliver something to the requesting CP, when a CMP wants to monitor a CP, and so on.

Let's look at some simple examples to see this in practice. Figure 5.6 shows us part of the CMP that we might have in a pharmaceutical company for clinical trials. We have shown just three of the potentially many threads that might make up such a process. They are all in the main role *Clinical Project Leader:*

- the thread that responds to a request for a new clinical trial;
- the thread that carries out regular monitoring of all the clinical trials currently in progress;
- the thread that responds to announcement of the completion of a clinical trial.

Spend a moment examining these three threads:

1. The top thread starts with a request for a new clinical trial from a requesting process. This process interaction takes the form of a role interaction between *Clinical Project Leader* and the 'anonymous' role *Requester*. If a clinical trial is to be done in-house, **Manage the flow of Clinical Trials** immediately activates its CP – **Handle a Clinical Trial** (Figure 5.7) – by instantiating that process's lead role, *Clinical Trial Leader.*

2. The middle thread involves an interaction with each of the current instances of **Handle a Clinical Trial** in the form an interaction with their lead role instances.

3. The bottom thread captures an interaction with an instance of **Handle a Clinical Trial** which is completing. This process interaction takes the form of a role interaction between *Clinical Trial*

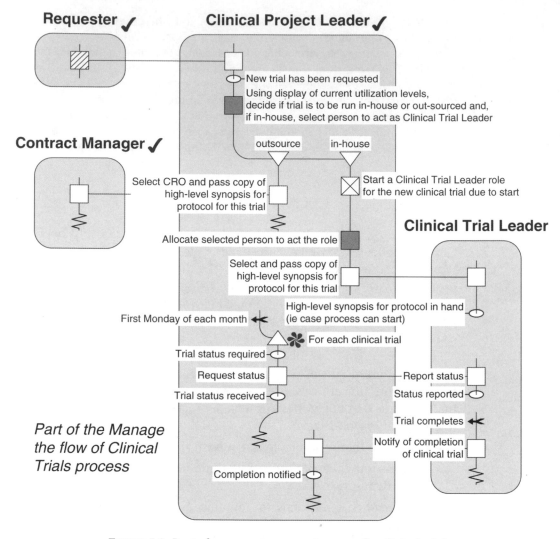

Requester ✓ **Clinical Project Leader** ✓

New trial has been requested

Using display of current utilization levels,
decide if trial is to be run in-house or out-sourced and,
if in-house, select person to act as Clinical Trial Leader

Contract Manager ✓

outsource   in-house

Select CRO and pass copy of
high-level synopsis for
protocol for this trial

Start a Clinical Trial Leader role
for the new clinical trial due to start

Clinical Trial Leader

Allocate selected person to act the role

Select and pass copy of
high-level synopsis for
protocol for this trial

High-level synopsis for protocol in hand
(ie case process can start)

First Monday of each month

For each clinical trial

Trial status required

Request status   Report status

Trial status received   Status reported

Trial completes

*Part of the Manage
the flow of Clinical
Trials process*

Notify of completion
of clinical trial

Completion notified

FIGURE 5.6 *Part of a case management process for clinical trials*

Leader in **Handle a Clinical Trial** and *Clinical Project Leader* in **Manage the flow of Clinical Trials**.

Figure 5.7 shows some of the CP **Handle a Clinical Trial** and three threads in particular within *Clinical Trial Leader.*

4. The top thread begins with the state that the lead role, *Clinical Trial Leader,* starts in – *High-level synopsis for protocol in hand* – and proceeds on its way.

5. The middle thread responds when necessary to a status request from the *Clinical Project Leader* in the CMP **Manage the flow of Clinical Trials** – this role interaction represents the process interaction in item 2 above.

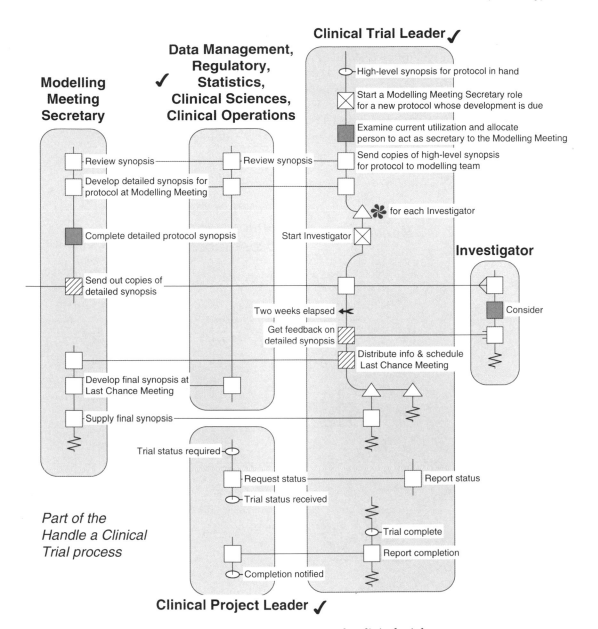

FIGURE 5.7 *Part of a case process for clinical trials*

6. The bottom thread is where the Clinical Trial Leader reports completion of the trial to **Manage the flow of Clinical Trials**; again the process interaction takes the form of a role interaction, corresponding to item 3 above.

We might also expect that the process – perhaps in the form of the Clinical Trial Leader – provides the results of the trial to the original requesting CP somewhere downstream.

Later still, next to the water cooler

Pupil: I'm a bit troubled by one aspect of the activation of CPs. When the CMP wants to activate the CP, it does it in the standard way: by instantiating the lead role of the CP. This implies that the lead role is a transient, probably abstract role, with an –ing name, such as *Batch Making*. But what actually happens of course is that a real, concrete role starts the work, such as *Plant Operative*. How do you reconcile these two things?

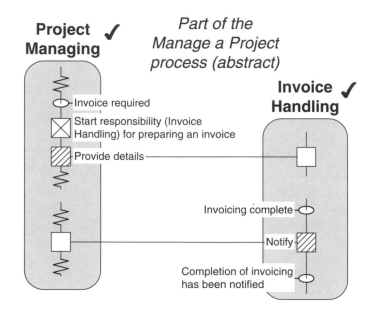

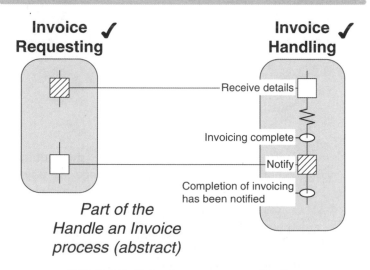

FIGURE 5.8 *Abstract case process activation*

Tutor:	A good point. When we want something done – an invoice prepared, a batch made, an examination paper marked – we are basically creating the responsibility for doing it. That's what instantiating a role is about. So I hope you can see that instantiating the (abstract) lead role does make sense.
Pupil:	Yes I can, but square that up with the concrete side of things for me.
Tutor:	You've answered the question: are we preparing a concrete model or an abstract model? If it's an abstract model we shall see something like Figure 5.8, in which we instantiate the lead role to create the responsibility.

If we want a concrete model we shall have something like Figure 5.9, in which we simply interact with the concrete role and fire up its main thread

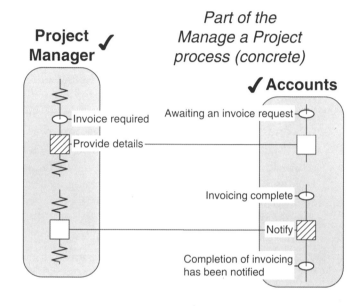

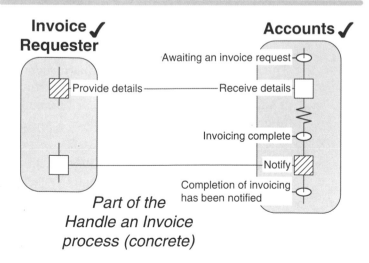

FIGURE 5.9 *Concrete case process interaction*

to deal with the new case – what we called a 'service interaction' in Chapter 2.

Which of these perspectives we choose depends on our reasons for modelling, as always. But I hope you can see that they both allow us to represent the concurrency that is possible with many cases in hand at once. In the abstract scheme (which I would prefer, all things being equal), each case has its own instance of the lead role. In the concrete scheme, each case has its own 'instance' of the thread of the main role.

THE CASE STRATEGY PROCESS

So far, we have seen how the CP deals with an individual case of a UOW and the CMP deals with the flux of instances of the CP or, as we put it, 'manages the flow of' those cases. Now suppose we walk into the building of the organization we are interested in and look around us. Some people will be engaged in dealing with cases, others will be managing the flow of cases. Have we accounted for all the activity in the building if we list all the CPs and their CMPs? Not yet. There is one more sort of activity going on and it involves people 'standing back' and taking the long view, the strategic view, of what is happening. The CPs are about front-line, coal-face activity; the CMPs are about managing that coal-face activity. *Case strategy processes* (CSPs) are about taking the strategic view of UOWs and driving the CPs and CMPs accordingly. A CSP has its CP and CMP as subject matter. It asks questions such as:

- What is happening inside our business that will affect my UOW and how it is dealt with?
  We are a water supply company and we have a big drive on to get more households on metered supplies. Meters need reading, and readings raise queries. How will the metering drive affect call rates at the Call Centre?

- What is happening outside our business that will affect it?
  The regulatory bodies in the pharmaceutical industry are taking greater interest in how we develop the software that we use in the R and D phase of our work. How will that affect the processes we use to develop that software?

- Is the nature of the UOW changing?
  Bank customers are increasingly prone to moving their accounts from one bank to another in response to changes in charges. How should our handling of customers change to accommodate this trend?

- Are the rates or volumes of our UOW changing?
  There are some significant changes going on in the types of courses that students sign on for. What does this mean for the processes that manage the courses that are suddenly finding themselves popular?

- What is the performance of our CP and CMP? Is it adequate and can it be improved?
  We have collected measurements of throughput and turnaround time for the processing of job applications. Do we need a different approach to scheduling in the CMP, combined with reorganized responsibilities in the CP?

- Are our CP and CMP actually being carried out in accordance with company procedures?
  The industry regulator takes a keen interest in the way our software is developed. Let's audit what is actually happening on the ground in our CP and CMP (instances) and ensure that we are up to the mark.

We collect these strategic considerations into the CSP. Typically, a CSP will have many threads, rather like a CMP. Some threads will be event-driven, triggered when something happens inside or outside the business that needs consideration and perhaps a response. Other threads will be calendar-driven and involve regular strategic reviews of business trends or industry trends. Either way, the outcome of the work of a CSP is likely to be changes or instructions to its CP and CMP.

Naming case strategy processes

As with the other two basic process types, we shall name CSPs in a stilted way as a starting point:

- **Maintain a strategic view of purchase orders.**
- **Maintain a strategic view of customer complaints.**
- **Maintain a strategic view of production batches.**
- **Maintain a strategic view of project reports.**
- **Maintain a strategic view of marketing campaigns.**
- **Maintain a strategic view of house purchases.**

The goal as ever is to focus our minds on the exact area of concern of this process, and to separate it from the concerns of the coal-face processes and their management processes.

KEY POINTS

The case strategy process takes the long-term internal and external view. It treats its case process and case management process as subject matter and might cause them to be changed.

SUMMARY

We can bring together the three basic process types in one diagram: Figure 5.10. The hypothesis in Riva is that any activity we observe when we walk into the organization's building will be part of a CP, a CMP or a CSP. This trinity of processes and their relationships give us the underpinning theory that we need to determine the organization's process architecture, as we shall see in the next chapter.

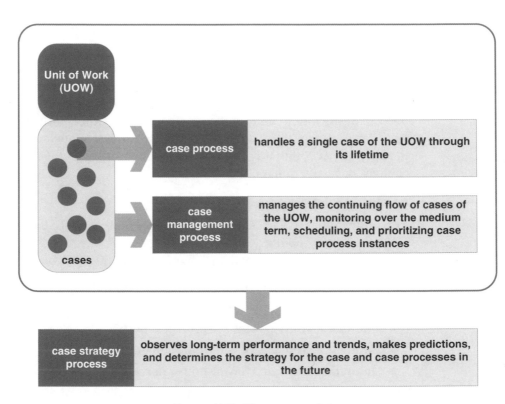

FIGURE 5.10 *The process trinity*

6 Preparing a process architecture

Describes how to construct the process architecture of an organization, a concept of central importance for any work with processes.

INTRODUCTION

A constant theme of the Riva approach is the central place of concurrency in organizational activity: the fact that when we walk into the organization's building we are surrounded by many instances of many processes all operating at the same time, some activating others, some interacting, and so on. At any one moment there is a network of interacting process instances at work. If we want to get our arms round all of that activity, to 'chunk' that activity into the constituent processes, how should we go about it? This chapter describes how to prepare the *process architecture* of the organization, the picture that says what process types there are in the organization and what their dynamic relationships are. Remember that I am not interested in static hierarchical decomposition – the world is not a neat hierarchy fixed in time – it is a dynamic network of interacting instances.

Getting this process architecture right is vital. A poor division of organizational activity into processes can easily lead to unnecessarily complex designs or models. There is a lesson to be learnt here from the world of software design: when we invent software modules we look for what are called high 'cohesion' and low 'coupling'. The idea is that each module should contain closely related activity: it is coherent in its content. The result of doing this is that the bandwidth of the relationships between it and other units is reduced: the system is loosely coupled. This in turn means that we keep the interconnections between modules to a minimum, thereby reducing complexity, and hence increasing maintainability and reliability. Software of course is synthetic: how it looks and how it is structured is entirely up to the software designer. A business process is typically not synthetic: it has developed organically and is the result of some design and a lot of chance. We cannot expect it to be tidy. But when we chunk the organizational activity we can expect to look for 'natural' or latent modularity: chunking that is there because it has to be, because the organization is in this particular business.

It is as if we are looking for natural fault lines in a rock when carving, or splitting wood with the grain rather than across it. When we model the real world, if we break the activity 'with the grain', we shall get fewer broken connections. We want to exploit any cohesion present in the real world. Our aim must be to find that natural grain.

If we are considering re-engineering, a bad initial division can at best obscure the possibility of a radical change to a process and at worst lead to the local optimization of processes at the expense of the organization. We also want to be able to distinguish between processes that are there because we are in a particular business and processes that are there because we have chosen to work in a particular way. The first group arise from what we shall call *essential UOWs*, and the latter from what we shall call *designed UOWs* – the latter are of course candidates for re-engineering.

People (especially those of us with software engineering backgrounds) have a natural desire to decompose a process into successively smaller subprocesses, in other words to draw up a hierarchical structure. But the world is rarely constructed in such a neat way and the processes in an organization are invariably connected in a network, rather than being contained in one another. The danger of thinking hierarchically cannot be overemphasized. To draw a diagram such as the one in Figure 6.1 is not to

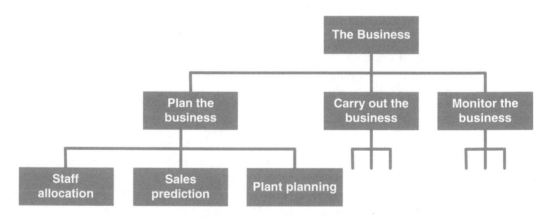

FIGURE 6.1 *Not a process architecture – more a random hacking*

prepare a process architecture. What does it mean to say, as the diagram implies, that one process is the 'sum' of three others? Where in the figure are the *dynamic* relationships between processes represented? It is, after all, the dynamic relationships that are crucial to an understanding of how an organization *works*. It is network dynamics that matter, not some arbitrary hierarchical composition.

It is equally all too easy to draw up a list of 'major' processes in an organization by looking for functional units in the organization and imagining that they in some sense represent 'top level processes', e.g. the 'Finance process', the 'Analytical Chemistry process', or the 'HR process'. These are all meaningless and probably misleading titles that must be resisted at all costs. If we change the organization's structure, its list of processes would change too, which would be absurd if it was still in the same business.

Figure 6.2 is an equally tempting 'chunking' of activity. At least in this picture there is a sense of dynamism, of the order of things. But it is

FIGURE 6.2 *Not a process architecture – more a list of silos*

hopelessly serial, and all too easily becomes just a list of silos.

Surely, what we are looking for is *a process architecture that is derived solely from an understanding of what business the organization is in*. We would like to say 'If the organization is in this business then it must have these processes.' 'This is the grain of the organization.' If the organization changes the business it is in, then we would expect the list of its processes to change appropriately. But if the business reorganizes itself, perhaps from a hierarchical to a matrix organization, why should the list of processes change? Clearly, how some of those processes are *done* might change, but not the list itself. If the business changes its culture from being production-focused to customer-focused, why would the list of processes change? If two departments are merged, why would the list of processes change?

We shall see that a Riva process architecture is an *invariant* for an organization that stays in the same business. This makes it a particularly secure place to start any process design, or improvement activity, or computer system design. Even if I am asked to look at what is perceived to be a single process, I insist on preparing a process architecture in order to ensure that we have the right chunking and that what appears to be one process is one process, rather than parts of several related processes arbitrarily lumped together. Never forget the axe.

Before we move on, let's remind ourselves of how we shall be using the word 'organization': an organization might be Jane and John, a project team, the HR Department, the sales force together with the production planning group, the Bristol branch, the four branches in Somerset, the entire company, or our market place and all its players. It's whatever we want to look at.

WHAT BUSINESS IS THIS ORGANIZATION IN?

Essential business entities

Our first step in constructing the process architecture is to *characterize the business that the organization is in*. We shall do this in terms of its *essential business entities* (EBEs).

In any business there are things – entities – that one cannot get away from. They are there simply because of the business the organization is in.
For instance:

- In a pharmaceutical R and D company we would recognize *Drug compound, Clinical trial, Assay,* and *Batch of raw compound*. As long as the pharmaceutical company is in the business of developing compounds that are subject to regulatory control, these things will exist: they are the *essence* of the business.

- If we are in the business of administering a modular programme in a university faculty, candidates for EBEs would be things like *Module, Award, Student assessment, External examiner, Curriculum* and *Appeal*. Take any one of these away and the business changes – or the faculty cannot do its allotted business.

- If we are a water utility, we shall think of things like *Customer, Supplied property, Meter, Asset, Job* and *Customer contact*.

- A firm of consulting engineers will have things like *Customer, Project, Market sector* and *Expertise area* on its list.

- A car repair shop will have *Customer, Appointment* and *Item of equipment*.

An EBE can be something physical and concrete such as a batch of drug compound (you can stick your finger in it in the barrel once it exists), or something rather abstract such as a clinical trial (you could see a clinical trial going on but you couldn't touch it), or something entirely abstract such as a request to change a clinical trial (e.g. 'I have decided to double the strength of the tablets'). The entirely abstract EBEs often prove to be very important, as we shall see.

(The older among us will hear echoes of Michael Jackson's JSD (Jackson, 1983). A JSD *real-world entity* must 'perform or suffer actions, in a significant time-ordering, exist in the real world . . . and be capable of being regarded as an individual.' Much of what follows has strong parallels with JSD's concepts.)

Sometimes, something that at first glance feels like an EBE is there only because of the way the organization has decided to do its business. How should we react if we find *Invoice* on the list of candidate EBEs? Someone in the room will argue that we can't survive unless we send out invoices; we shall run out of money, and go bankrupt: 'They're essential – *Invoice* must be an EBE'. For a car manufacturer, *Invoice* is not an essential entity;

it is a *designed* business entity. The car manufacturer is not in the business of invoices. On the other hand, *Payment* is essential and not designed. Invoices might be the way the manufacturer has decided to obtain payment: we must get paid, but how is a matter of choice. However, for the Invoice Handling Department – which *is* in the business of handling invoices – *Invoice* is certainly an EBE.

An EBE is called 'essential' because it is part of the essence of the business. Unless we are in the Invoicing Department, invoices will not be what we are about: we are not in business to handle invoices, invoices are not our subject matter, they are not what we are in business for. They are not 'essential' in that meaning of the word. And, of course, we can also argue that they are not essential in the other meaning, that 'we have to have them'. They are, after all, only a way of requesting payment, and we can think of many other ways of doing that. The fact is that we have *chosen* invoices as the way we will ask for payment. Invoices are certainly business entities for us, but they are designed business entities, and not essential business entities – they are there because of the way we choose to do our business, rather than because they fundamentally characterize our business.

This leaves us with a modelling decision. How should we handle designed business entities? The answer is to consider them, whilst noting that they are designed entities, and to replace them – where possible – by the EBEs they stand for or implement.

Think of the list of EBEs as a searchlight that illuminates the part of the world we are interested in. We can focus in on the HR Department, or expand out to the entire company. We can illuminate both the pharmaceutical drug company and its partner companies in clinical research, or we can restrict our focus to the drug company alone. We can concentrate on some part of our 'internal' world, or we can extend our view to cover what our customers are concerned with. Indeed, whilst we might start by thinking about a particular organization and then prepare a list of its EBEs, we shall invariably find ourselves revisiting exactly what 'organization' we want to cover and then redrawing the boundaries differently in order to answer the questions we want to answer.

EBEs are a powerful way of focusing our thoughts on the things that matter, and avoiding analysis paralysis.

> **KEY POINTS**
>
> The process architecture must capture the network of concurrent activity in the organization.
>
> It should be derived solely from an understanding of what business the organization is in.
>
> We start the analysis and characterization of that business by examining its EBEs.
>
> An EBE is an entity that is the essence of the organization's business.
>
> An EBE is part of the subject-matter of the organization.

Finding the EBEs

Since the EBEs are the subject matter of the organization, anyone who knows what business the organization is in could – in principle – quickly give you a list of EBEs. So let's get half a dozen of those people in the room to brainstorm the list.

In true brainstorming fashion, the list will quickly build to thirty, forty, fifty, a hundred candidate EBEs. We can prompt suggestions with questions such as:

- **What do we make?**
  Cars, packs of biscuits, radios, furniture, bottled drinks, . . .
- **What do we sell?**
  Cars, palettes of packs of biscuits, water, electricity, insurance policies, items of furniture, packs of tablets . . .
- **What product lines do we have?**
  These models of cars, these ranges of biscuits, these designs of furniture, . . .
- **What services do we offer?**
  Giving roadside assistance for a vehicle breakdown, responding to an emergency call, answering a customer complaint, . . .
- **What service lines do we have?**
  These types of insurance policy, these levels of maintenance service, these levels of call-out service, these types of financial portfolio management, . . .
- **What things can we simply not get away from?**
  We are developing pharmaceutical drugs, so we cannot get away from the regulatory authorities.

  We build aero engines, so we cannot get away from the safety regulator.

  We are a quoted company, so we have shareholders.

  We have staff members.

  Company policy requires us to follow certain quality standards.
- **Who are our external customers?**

Car buyers, car dealers, car wholesalers, fleet car buyers, ...

- **Who are our internal customers?**
  Researchers, project managers, staff members, the Board, ...

- **Are there things that our customers have, or want, or do, that might be EBEs for us?**
  Complaints, purchases, overdrafts, accounts, loans, loyalty cards, ...

- **What things do we think differentiate our organization from others in the same business?**
  Our quality focus, our culture, our expertize, our prices, our brands, our customer focus.

- **What sorts of things do we deal with day in, day out?**
  Car engines, flour suppliers, drilling machines, power stations, customer complaints, machine failures, quality standards.

- **What events in the 'outside world', the world outside our organization, do we need to respond to?**
  Power failures, drain collapses, customer complaints, significant share price changes, new financial years, ...

- **What business entities are listed in our corporate data model?**
- **What things do our information systems keep information on?**

Once the flow has dried up, we revisit each item on the list and apply a number of filters that test whether it is truly *an entity that is the essence of the business*. I like to keep the entire list of candidate EBEs as we move forward, whilst bracketing out – rather than deleting – those that don't get through the filters, and noting the reason why each has been bracketed. This is going to be an iterative process and we are certain to revisit this list later, perhaps reinstating a candidate, or adding new candidates. So it is a good idea to keep all that work and all those decisions as the work proceeds. Here are the filters:

- Since these are all supposed to be *entities*, test each by putting the word 'a' or 'the' in front of each suggestion. If it doesn't make sense, bracket it and think again: are there any other entities that are suggested and that do pass the test?
  This can cause consternation in some situations. If we are a water supply company, then 'water' surely must appear on the EBE list, yet you can't talk about 'a water', or, indeed, 'an electricity' if we were an electricity supply company. Have faith – bracket it and keep going.

- Bracket any designed entities.
  Invoices may be the meat and drink of the Invoicing Department and hence an EBE for them. But for the company as a whole, they are not: the organization is not in business to issue invoices. Invoices, for the organization, are just a way of obtaining payment.

- Bracket entities that are simply roles, and which are not 'of the essence' of the business.

  The Accounts Department is an entity, and it's about the business. Someone (possibly from the Accounts Department) will say 'This company wouldn't function without the Accounts Department – they're essential.' Well yes, they are essential, but not 'of the essence': we are not in business to do accounts, we make cars. Accounts play an important role. But they are not an EBE, though they have a role.

WHICH EBES REPRESENT WORK FOR THE ORGANIZATION?

Filtering off units of work

After the first filters have been applied to our brainstormed list of candidate EBEs, we shall have reduced the list to perhaps half. The items remaining are true EBEs. These are the things the business has as its subject matter. They define and characterize the organization.

Our second step is to decide which of these are entities that have lifetimes during which we must look after them. These are our units of work. Essential business entities can be *essential UOWs*, and designed business entities can be *designed UOWs*.

For instance, a clinical trial is a UOW for an R and D pharmaceutical company: it starts, proceeds and stops, and we must look after that life. A drug compound is a UOW: it is invented, tested and developed, taken to market, and finally withdrawn, and we must look after that.

We need a further set of filters to help us whittle the EBEs down to just those that are UOWs:

- Bracket EBEs that are clearly not UOWs for us.

  A *Purchase* of a theatre ticket would be an EBE for a Box Office and a UOW. A *Ticket* might be on our list of candidate EBEs, but we can bracket it because the ticket itself does not have a separate lifetime of its own of interest to us: we don't care how it is designed, printed and distributed. It's just a mechanism we use in the contract – it *stands for* a successful purchase and the right to occupy a seat at a performance. ('Ah, did we have "Performance" on the list? Is that a UOW for the Box Office?')

- Bracket EBEs that are not UOWs for us, even if they are for someone else.

  A 'quality standard' clearly has a lifetime. If we are the central Quality Management Group, with responsibility for the quality standards that make up the Quality Management System, then quality standards are our meat and drink and that lifetime would certainly represent a unit of work for us: we shall decide on the need for a standard, draft it, have it reviewed, approve it, distribute it, make changes to it, and finally withdraw it. But for somebody who is only required to use quality standards, they may be an EBE but they are not a unit of work. Standards will only feature as controls in their processes, and will not be the subject matter of their processes.

- Bracket EBEs that are only roles that play a part in processes.
  The Safety Regulator is an EBE of, say, a company producing railway signalling systems. But there is no sense in which the *lifetime* of the Safety Regulator is something that the company concerns itself with: it does not have to look after the Safety Regulator in that sense. Clearly though, the Regulator will play a role in many of the company's processes, and we will expect to see it as a grey box on RADs.
  It's quite typical for the list of candidate EBEs to contain all sorts of job titles and posts in the organization: CEO, Project Manager, Salesperson and so on. True, the business could perhaps not operate effectively without the CEO but the CEO does not have a lifetime that we need to handle. If we cannot imagine having a process to look after that role during its lifetime then we bracket it.

- Bracket any EBE that is only part of another EBE and does not have a separate lifetime of its own.
  If we manufacture DVDs, then *DVD* is probably an EBE and a UOW. But *Jewel case* is probably not a UOW for us. Yes, each compact disc we make goes into a jewel case, but we don't care about the lifetime of that case: we just buy them in. The disc itself does have a lifetime however, from the moment it is a lump of molten plastic to the point where it is capable of being played in a DVD player.

Some EBEs might (or might not) turn out not to be UOWs themselves, but might point to other UOWs. This is commonly the case with collections of things that form another thing:

- As an electricity distribution company, we regard (the) *Transmission System* as an EBE – without it we are not in business – and there is a very real sense in which it has an (unending) lifetime that is of interest to us. But perhaps more interestingly we realize that the *Transmission System* is a collection of assets and *Asset* should be on the list too. So we add *Asset* to the list and show both as UOWs.

- *Our expertize* is in the list. Is it helpful to think of the lifetime of (the) expertize that we have? Probably not. However, *Our expertize* is made up of a number of individual expertizes, and each of these will be chosen, developed, fostered, and perhaps finally dropped. So *Expertize*, on the other hand, does have a lifetime of interest to the organization: that is a UOW.

- As a pharmaceutical development company, we are very concerned to have a good 'pipeline', i.e. a good stream of potential new drugs going through research and development. That *Pipeline* will appear on our EBE candidates list. But so will *Drug compound*. We can view both as

EBEs and also as UOWs. Of course, we have only one pipeline, but that pipeline is composed of a number of compounds.

Finding 'unseen' UOWs

As in any requirements-gathering exercise the question arises: 'How do we know when we have finished? How do we know when we have everything we should have?' We have a number of ways of checking for UOWs that we might have missed:

- Examine the names of departments and groups.
  Individual departments often exist to deal with one sort of UOW: for instance, we might observe that the Analytical Chemistry Department deals with assays. Is 'Assay' an EBE and a UOW perhaps for the organization? The Emergency Response Team deals with ... emergencies. The Help Desk deals with fault reports. We need to be careful when finding UOWs this way: they might well be designed UOWs.

- Put the words 'Change to' in front of each candidate UOW and see if it creates another UOW.
  At one pharmaceutical company we found people who thought their unit of work was dealing with requests from clinicians for the patient packs (of drugs) for clinical trials. When we looked, we found that they did indeed do that and had a UOW called *Request for supplies for a clinical trial*. But additionally – and very importantly – they spent a great deal of time dealing with changes of mind from clinicians, so they had a further unit of work which was the *Change to request for supplies for a clinical trial*: a change arrived, it had to be dealt with, and finally had to be incorporated into readjusted schedules. For each original request there might be several changes.

- Put the words 'Collection of' in front of each candidate UOW and see if it creates another UOW.
  Is there a sense in which the collection of things has its own existence? Like the portfolio of drug compounds in a pharmaceutical company, or the product range of a software application company, or the publisher's list of books in print. A publisher doesn't just deal with titles in isolation: it will build a list of a particular character and content. In other words, the list has its own existence and lifetime, apart from the titles that make it up.

> **KEY POINTS**
> A UOW is an EBE that has a lifetime during which we must look after it.
> These are essential UOWs.
> Other UOWs arise from designed business entities rather than from
> EBEs. These are designed UOWs.
> Some UOWs are collections of other things or changes to other things.

Tips for the process architecture brainstorming

- Work on flipcharts so that you have a full record of what happened. Whiteboards invite rubbing out and reduce the ability to backtrack earlier decisions. This also helps people to remember how they got to where they got and why.

- Treat the brainstorming of candidate EBEs as true brainstorming: write them down as they are shouted out, without discussion. The one exception I make to this rule is to apply the 'a/the' rule to check that something is an entity.

- When starting the filtering for real EBEs, start with easy ones to help people get their heads round the idea. If we make cars, then *Car* is a true EBE. If we design cars, then *Car design* is a true EBE. But *CEO* is clearly just a role.

- Do the same when filtering for EBEs that are also UOWs. If we make cars, then *Car* is a UOW. If we are in a regulated industry, the *Regulator* is a true EBE but not a UOW.

- Draft a sentence describing each UOW, in particular to capture its scope. This is an aide memoire for future work and future readers.

WHAT ARE THE DYNAMIC RELATIONSHIPS BETWEEN UOWS?

Our list has now been whittled down to those true EBEs which are also UOWs. The next step is to examine the relationships between those UOWs.

Now, we can think of all sorts of possible relationships between UOWs: the annual product portfolio review process (**Carry out the annual review of the portfolio**) perhaps needs information from the process for collating the annual accounts (**Prepare the annual accounts**). That's an information relationship; these two processes will therefore need to interact to exchange information. The question is 'which relationships are interesting?' Because we shall be interested in the *dynamic* relationships between the processes, we shall want to concentrate on the *dynamic* relationships between the UOWs.

When we were examining the different dynamic relationships that can exist between processes, we saw that relationships arise when some UOWs 'need' other UOWs. For instance, candidate drug compounds require clinical trials to be carried out. During the development of the compound many clinical trials will be required. There will be times when several clinical trials will be in progress for the same compound. Every clinical trial requires patients. They have to be recruited; they have to be given the drug, a placebo, or a comparative drug; and they have to be monitored and recorded. At any time during the trial many patients will be participating in it. In order for a patient to participate in a clinical trial 'patient packs' have to be prepared that contain the doses of whatever they are receiving for the trial. Since there are many patients in a trial – potentially thousands – there will be a need for many patient packs.

Compound, *Clinical trial*, *Patient* and *Patient pack* are all UOWs in the pharmaceutical R and D company. And there are important dynamic relationships between them. We summarize this sort of relationship with the neutral word 'generates':

- A candidate drug compound generates clinical trials.
- A clinical trial generates patients.
- A patient generates patient packs.
- A project generates work packages.
- A customer generates customer orders.
- A staff member generates expense claims.
- An order generates batches.

The word 'generates' perhaps jars in some cases, but I use it as a catch-all word covering concepts such as 'need', 'require', 'call for' and 'activate'. And when I say that '*A* generates *B*' I mean 'during the lifetime of a case of UOW *A*, cases of UOW *B* are needed/called for/... ' So, during the lifetime of a clinical trial, patients are needed. During the lifetime of a customer, orders are placed by that customer. During the lifetime of the portfolio, (new) products are considered. And so on.

In some cases, the lifetimes of the cases of UOW *B* are all contained within the lifetime of the case of UOW *A*: clinical trials must have been completed before development of the compound can be completed. Sometimes the generated cases of *B* live on, after the generating case of *A* has finished: a project will generate invoices to the client, and their handling can linger on after the project itself has finished. We might allow the project to close even though the processing of its invoices continues after the project is declared to be finished.

The *cardinality* of the relationship between two UOWs can vary: each case of *A* generates exactly one case of *B*, or each case of *A* generates none, one, or many cases of *B*. One clinical trial generates one clinical trial report. One clinical trial generates many patients.

Our third step in building the process architecture is to go through the filtered list of UOWs and draw out these dynamic relationships, the 'generates' relationships. We construct a *UOW diagram* in which we show the UOWs and the dynamic relationships between them; we name each relationship and identify its cardinality (one-one or one-many). For example, 'A Compound requires *many* Clinical trials'. We use the following conventions:

- Each UOW is shown as a hexagon containing the name of the UOW in the singular.
- Each 'generates' relationship is shown as an arrow from the generating UOW to the generated UOW, and appropriately labelled.
- Where a UOW is generated by an agent outside the organisation we are concerned with, we show the arrow coming from a cloud suggesting 'The Outside World'.

Figure 6.3 gives an example of the sort of UOW diagram we might produce. It shows the UOWs that characterize the administration of the teaching programme in a university faculty and their dynamic relationships.

KEY POINTS

The UOW diagram shows only dynamic relationships between UOWs.
In particular, it shows only 'generates' relationships, where one type of UOW arises because of and during the lifetime of another.
No other relationships are drawn on the diagram.

Later, next to the water-cooler

Pupil:	I'm worried that in your UOW diagram for the university faculty some UOWs are not connected to others. *Student,* for example, is just sitting on its own. And *Teaching quality event* and some others are generated 'out of the blue' but don't generate anything else.
Tutor:	Do not be alarmed by this! There is absolutely no requirement for every UOW to be connected to some other UOW. Remember that we are only looking at this stage for those dynamic 'generates' relationships. Not everything generates something else or is generated by something else!
Pupil:	Well, I'm tempted to suggest that there must be some relationship between the *Student* and a *Student assessment.*
Tutor:	It's very important at this stage to draw only the dynamic relationships. Do not be tempted to think of other – possibly interesting – relationships. A Student assessment is generated from the lifetime of a run of a module. Yes, it concerns the student but it isn't the lifetime of the student that generates the assessment. This is a subtle but important point.
Pupil:	How come students appear out of thin air?
Tutor:	As far as this organization – the Faculty Administration – is concerned, they do! We've turned the searchlight away from student recruitment towards the teaching side of the faculty.

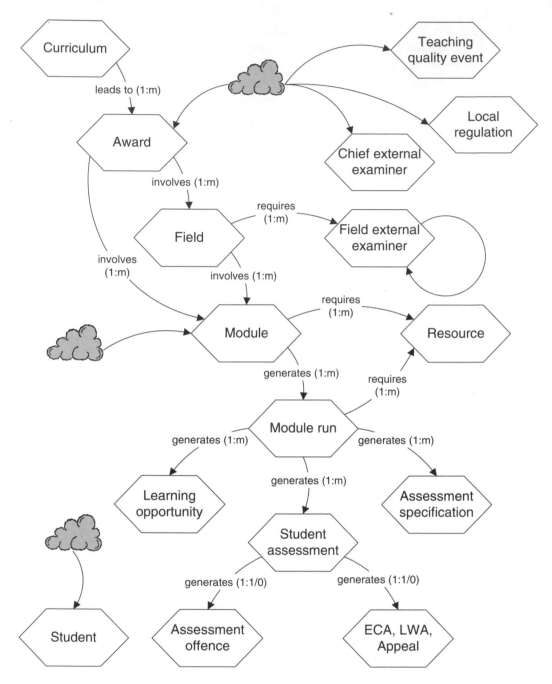

FIGURE 6.3 *The UOW diagram for a university faculty administration*

PRODUCING THE FIRST-CUT PROCESS ARCHITECTURE

We started this chapter with the assertion that a true process architecture should be an invariant of the organization, determined only by the business the organization is in. So far, we have developed a characterization of that business in terms of a set of essential UOWs and their dynamic relationships, all drawn up on the UOW diagram. The next step – producing a 'first cut' of the process architecture – is entirely mechanical and uses the ideas we developed in the previous two chapters.

Firstly, we hypothesize that for each UOW on the UOW diagram there are three processes: its CP, its CMP, and its CSP. So, for the UOW *Customer call*, we know that **Handle a customer call, Manage the flow of customer calls** and **Maintain a strategic view of customer calls** will all appear somewhere in our process architecture. By definition, if *Customer call* is a UOW, we have to look after it during its life time: that is what the CP does – it looks after one instance. Since there will potentially be many in progress at any one moment we need to handle that flow: that is what the CMP does. Since this thing is either of the essence of the business or important enough to be designed, somewhere we need to take a strategic view of it: that is what the CSP does.

Secondly, we hypothesize that each 'generates' relationship between two UOWs in the UOW diagram can be translated into relationships between the corresponding processes. Here we use the results of Chapter 5. We examine each 'generates' relationship and decide if it is a task force or a service function relationship.

If *A* generates *B* and the relationship is a service relationship, then the translation rule in Figure 6.4 is used.

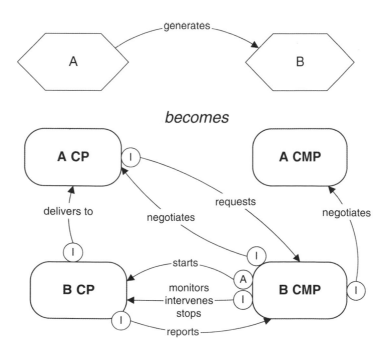

FIGURE 6.4 *Translating a service relationship between UOWs into processes*

If the relationship between *A* and *B* is a task force relationship, then the translation rule in Figure 6.5 is used.

Note that in both cases we have drawn the full set of relationships between CP and CMP: starting, monitoring, intervening, stopping, reporting and negotiating.

Figure 6.6 shows the UOW diagram we might draw for the area in a company to do with looking after the system that supports the business requirement. There is one business requirement, which has a lifetime that is looked after and during which changes arise. Matching it is one system, which also has its lifetime. During the lifetime of the system, new releases are made of it. Looking after it also requires new work products to be generated. And so on. Spend a moment understanding the dynamics of this organization.

Knowing how to deal with the two relationships of service function and task force we can, quite mechanically, transform the UOW model of Figure 6.6 into the first-cut process architecture in Figure 6.7. We have only labelled the 'encapsulates' relationships, as all the others should now be obvious from context.

(We shall generally omit CSPs from the process architecture unless they are of specific interest for our purpose.)

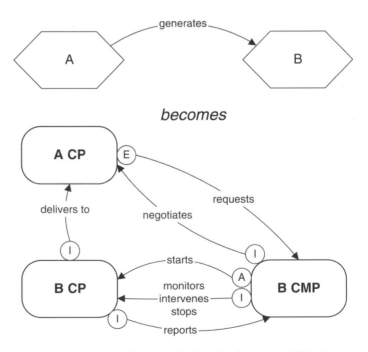

FIGURE 6.5 *Translating a task force relationship between UOWs into processes*

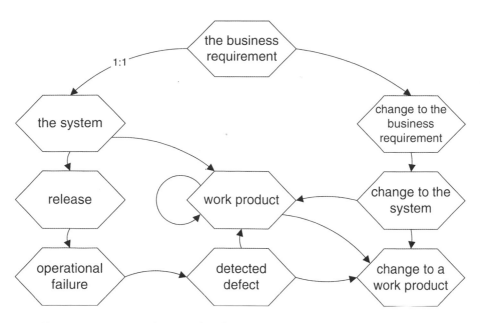

FIGURE 6.6 *A UOW diagram for the IS support to the business requirement*

Pupil:	It's complicated – should I be worried?
Tutor:	Ask anyone in a company looking after computer systems supporting business requirements and tell them it's simple! That said, this is the first-cut architecture. There are some reductions that we can make. But take careful note: they are reductions that mirror the real world. They are definitely *not* simplifications to make the diagram look simpler! That would do us no service at all.
Pupil:	OK, I can accept that. But people will surely be put off by the fact that the picture is a mass of blobs and arrows.
Tutor:	I'll put what you said another way: people will surely be put off by the fact that the business operates a complex network of interconnected processes. Whether or not they are put off, it is true! And we shall not get to answers if our pictures are lies or over-simplifications.

PRODUCING THE SECOND-CUT PROCESS ARCHITECTURE

That last step was purely mechanical. The resulting process architecture represents – in a sense – the most we can expect to find in the way of processes for the units of work we identified. In practice, it often shows more than exist. Let's explore some of the ways that we can reduce the first-cut process architecture to produce the second-cut process architecture.

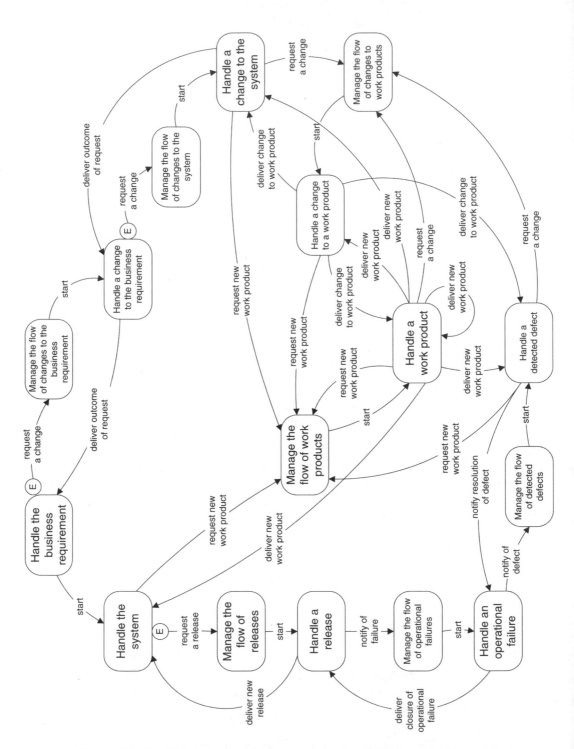

FIGURE 6.7 *The first-cut process architecture for the UOW model in Figure 6.6*

Folding a task force CMP into the requesting CP

Where a task force relationship has been transformed, and the CMP that receives requests is shown as encapsulated in the requesting CP, we can decide to fold the CMP into the requesting CP, particularly if it is trivial or near trivial. Why does this make sense? We know that the things being managed by the CMP are only requested by that CP – otherwise it would be a service function. So, it's plain that we can consider case management to be part of the requesting CP.

As an example, the fragment of first-cut process architecture on the left of Figure 6.8 is folded to give the fragment on the right. This arose from the

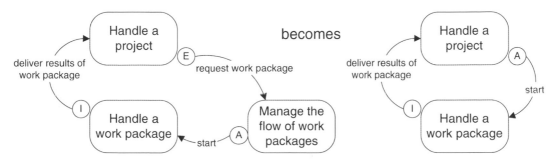

FIGURE 6.8 *Folding a CMP into the requesting CP*

UOW relationship 'A Project generates Work Packages' and a recognition that that relationship is a task force relationship: the project sets work packages going under its own initiative, rather than going to an 'outside' service that does work packages. Put another way, the **Handle a project** CP does its own case management for work packages.

It is vital to remember that when we fold the CMP into the requesting CP, we are *not* saying that the CMP doesn't exist, or that there is no case management to be done. We are saying that it sits within the requesting CP and is best modelled there. We are entirely at liberty to leave it as a separate process if we think that our purpose will be better served by doing that.

Dealing with 1:1 'generates' relationships

In some cases, we will have on our UOW diagram one instance of an *A* generating precisely one instance of a *B*, what we shall call a 1:1 relationship. For instance, we might have drafted the following:

- A Customer Purchase generates one Invoice.
- A Clinical Trial generates one CRF Design.
- A Draft Document generates one Approved Document.
- A Job Applicant generates one Staff Member.

There are some subtleties here.

Suppose that in our organization one *Customer purchase* generates precisely one *Invoice*. Assuming that invoicing is done, for all sorts of reasons, as a service within the organization, we shall want to keep the processes arising from *Customer purchase* and *Invoice* separate. We shall need the separate **Manage the flow of invoices** to sort out priorities between invoices from the different sources, i.e. from CPs other than **Handle a Customer purchase**. So the handling of invoices is a service function and we must leave the full set of CPs and CMPs on the second-cut architecture.

Suppose we are the Clinical Trials Division of a pharmaceutical drug development company. The Case Report Form (CRF) is a large document that records the life-history of a single patient taking part in a clinical trial. For each clinical trial, a CRF is specially designed to take into account the particular requirements of that trial. In Riva terms, a *Clinical trial* generates just one *CRF design*. Do we still need both UOWs? Yes, we probably do. The CRF design has its own life that runs alongside the trial's: it is drafted, reviewed, approved, copied, amended, and so on. However, since there is only one CRF design for one clinical trial, and its lifetime is intimately related to the trial and not passed off to someone else to look after, will we need case management for CRF designs? Yes, we probably do: there might only be one CRF design for each clinical trial, but there are many clinical trials and hence many CRF designs, and if we provide the development and maintenance of CRF designs as a service then we shall need both **Manage the flow of CRF designs** and **Handle a CRF design** in the process architecture.

A *Draft document* is a document in a particular state: draft. After it has been through certain processing it might change its state and become an *Approved document*. What happened there? Perhaps it would make more sense to say that a *Document* changes state, and hence to replace the two UOWs by one. The first half of **Handle a Document** would then be about the lifetime of a document in the draft state and the second half would be about a document in the approved state. We would not distinguish between case management of draft documents and case management of approved documents, contenting ourselves with **Manage the flow of Documents**.

Suppose we have a UOW called *Job applicant* and another called *Staff member*. In some cases, job applicants become staff members. In some cases they don't. Job applicants sometimes 'turn into' staff members. Strictly speaking, *Job applicant* is in a '1:0/1' relationship with *Staff member*. Following the *Document* analogy, we would try to find a single UOW of which *Job applicant* and *Staff member* were just different states. Hard. Moreover, the lifetime of a job applicant is quite different from the lifetime of a staff member – quite different things happen to you – and it feels better to keep them separate and hence with their own CPs. Also, the case management of job applicants is a wholly different thing from the case management of staff members. So, we shall find the full set of processes in the process architecture – **Handle a Job applicant, Manage the flow of Job applicants, Handle a Staff member**, and **Manage the flow of Staff members** – and at some point **Handle a Job applicant** will request a new staff member from **Manage the flow of Staff members**. That is the moment of recruitment.

Dealing with delivery interactions and delivery chains

The standard transformation always produces a 'delivers' interaction from the requested case to the requesting case. When a project generates a work package, the work package is assumed to deliver something back to the project. And in this situation that is probably what happens in reality. But in other situations the 'generates' relationship is more 'fire and forget'. So once we have the first-cut architecture in front of us, another validation we can carry out is to examine each 'delivers' interaction and ask 'Does this happen in reality? Does anything really get delivered?' If the answer is 'No', then we can delete that interaction.

Let's look at another situation. Suppose *A* generates *B* generates *C* generates *D* on the UOW diagram. This will yield the first-cut process architecture shown in the top half of Figure 6.9. Note how the chain of delivery is from *D CP* to *C CP* to *B CP* to *A CP*. It is always worth thinking this through and comparing it with reality. We often find that in practice

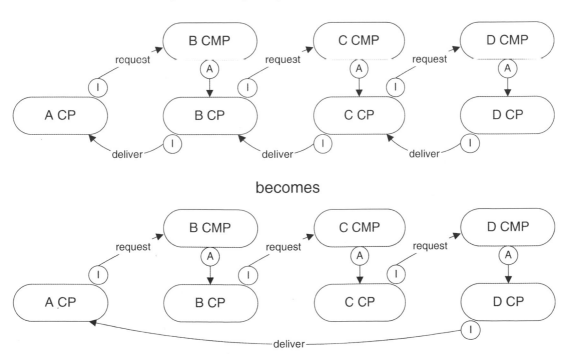

FIGURE 6.9 *Some delivery chains can be short-circuited*

this delivery chain is short-circuited and that the actual relationships are as shown in the bottom half of the figure.

We shall see an example of this in a case study later.

Dealing with collections

We saw earlier how our UOW analysis might turn up a UOW that is actually a collection of another UOW. For instance, a *Programme* is a collection of *Projects*; a *Project* is a collection of *Work Packages*; a *Transmission System* is a collection of *Assets*; a *Product Portfolio* is a collection of *Products*; and so on. In other words, during the analysis we have decided that not only does an individual project have a lifetime that needs handling, but a collection of projects – a programme – also has its own lifetime that needs handling. We will probably also have identified that (with our special use of the word 'generates'):

- A Transmission System generates Assets.
- A Project generates Work Packages.
- A Product Portfolio generates Products.

Let's take the first of these. It will produce the snippet of process architecture shown in Figure 6.10. Let's think what the process **Manage the**

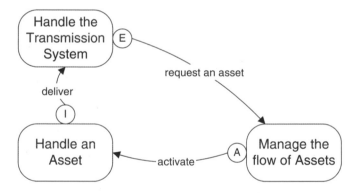

FIGURE 6.10 *UOWs, collections, and their processes*

flow of Assets is likely to be about. The more we think about it, the more we suspect that managing the flow of assets is actually part of what handling a transmission system is all about. So the CMP for the atomic object is the same as – or at least is contained within – the CP for the collection, and once again we are likely to fold the CMP into the collection's CP.

As another example, let's look again at the way in which a pharmaceutical development company runs the business of getting new chemical compounds into the drug market. We can identify an obvious unit of work: the *Candidate drug*. Each candidate drug will have its own life history: it starts as one of thousands of compounds that have been successfully screened, and it proceeds through a variety of ever larger trials; at the same time, there is a parallel development of the chemical process by which it will finally be manufactured in bulk. If the candidate drug proves safe and efficacious it will reach the market. This 'Molecule to market' process will involve many scientific groups that deal with the candidate drug: pharmaceutical sciences, clinical trials, manufacturing process scale-up, analytical chemistry, health and safety, quality assurance, and the regulatory group.

At any one moment the company will have many drug candidates at different stages. Such an organization is in fact a 'case pipeline', designed to get the successful candidates through to the market place in the shortest possible time and to weed out the unsuccessful candidates as soon as possible. (So few candidates typically make it to market that perhaps the process should be called 'Molecule to reject bin'.) The pipeline can be considered a UOW in its own right: it has a (never-ending) lifetime and must be looked after. Clearly, the pipeline is the set of candidate drugs. At any one moment there will be one pipeline containing many candidate drugs. Indeed, the pipeline generates candidate drugs, so we shall expect to find the three processes shown in Figure 6.11.

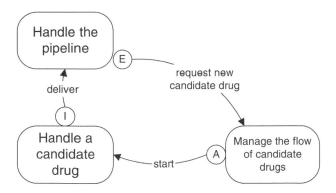

FIGURE 6.11 *The first-cut processes for a pipeline and its candidate drugs*

Now, requests for candidate drugs do not come from anywhere else other than the pipeline, and the pipeline *is* the set of candidate drugs. So a task force relationship is appropriate between **Handle the pipeline** and **Manage the flow of candidate drugs** (notice the E-relationship). This suggests that we can fold **Manage the flow of candidate drugs** into **Handle the pipeline**. This will make immediate sense: if we watch **Handle the pipeline** at work we will see that it is amongst other things about managing the flow of candidate drugs. The result will be Figure 6.12, but note that we are never obliged to fold the CMP away like this: it is a modelling decision to be made for each situation.

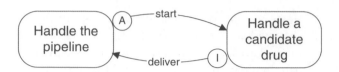

FIGURE 6.12 *The reduced processes for a pipeline and its candidate drugs*

Dealing with empty CMPs

It is not uncommon for a CMP to be ... empty: some things simply don't need or don't get case management. As a trivial example, we know in Figure 6.10 that there is only one Transmission System – there will be no process called **Manage the flow of Transmission Systems** and there will only be one instance of **Handle the Transmission System**.

Perhaps Jill deals with invoices. If you want an invoice, you walk up to her desk and leave an invoice request on it. She picks up invoice requests from her desk at random and works on them, sometimes working on several at once. The process **Manage the flow of invoice requests** simply does not exist. For our second-cut process architecture we might therefore remove it: any requesting CPs will simply activate **Handle an invoice request** itself, rather than asking a **Manage the flow of invoice requests** to do it. Dropping your request on her desk is all that needs to be done.

But then Jill decides to organize herself better: she now has an in-tray where you put your invoice request. It's annoying, but she simply takes the top request off the pile, deals with it, takes the next one off the top, and so on. Last on, first off. Now there is a **Manage the flow of invoice requests** process, and it has a serious effect on the variability of the time it takes for an invoice request to be dealt with.

From first-cut to second-cut architecture

We are now in a position to reduce the first-cut architecture in Figure 6.7 to the more realistic second-cut architecture in Figure 6.13. You can see that much of the reduction (a word I prefer to 'simplification') has occurred around **Handle the system** – not surprisingly as there is only one and hence it does a great deal of case management itself. Such reductions can't be made for the 'smaller' UOWs – e.g. *Work product* and *Change to work product* – where requests for them come from all sides and the flow is managed as a service. We have also short-circuited a couple of delivery chains to reflect reality.

Designed business entities and UOWs

When we analysed our candidate EBEs looking for UOWs, we bracketed out designed entities on our list of candidate EBEs and so they were not carried forward onto the UOW diagram. Our reason was that, if we want a process architecture that is solely based on the business the organization is

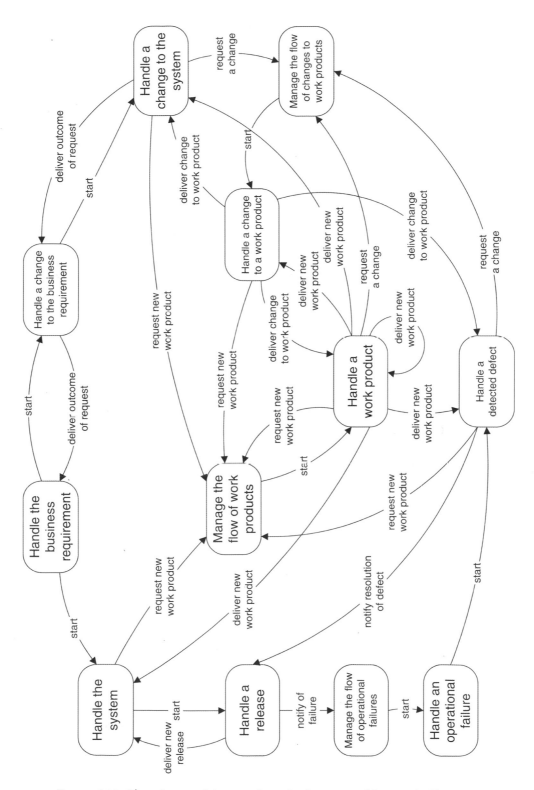

FIGURE 6.13 *Second-cut architecture from the first-cut architecture in Figure 6.7*

in and is independent of how it chooses to do its business, then we must drop them from the list. The result is that our process architecture really does concentrate solely on those processes that must exist because we are in the business we are in. There are no processes that are there simply because of some decision about *how* we should do our business.

This gives our process architecture a 'purity' that can be very useful in many situations. If we want to re-engineer then it is important not to be blinded by current mechanisms.

On the other hand, if we are building an 'as-is' view of the organization, then we shall want to see processes in the architecture that *are* there because that is the way we choose to do our business. So we can take the brackets off the designed UOWs concerned and let them generate processes in the architecture.

KEY POINTS

To build a process architecture:
1. Identify the EBEs.
2. Use the filters to extract the UOWs.
3. Map the UOW relationships and add their type and cardinality.
4. Transform the UOW diagram into the first-cut process architecture using the two standard transformations.
5. Consider folding some processes, especially where task force relationships are involved.
6. Put the resulting architecture against the world to validate the relationships.
7. Restore designed UOWs, if appropriate.
8. Make any further reductions and finalize the second-cut process architecture.

Other process interactions

When constructing the process architecture we deduced the principal interactions between processes: the request from a CP to a CMP, the delivery from a CP to a requesting process etc. There will of course be others: when we get into detail about how processes operate – how we have chosen to do our business – we shall see that there are other interactions between processes. But since they are about how we do our business, we would rather not have them on our 'pure' process architecture, an architecture that we want to be re-engineering-proof. They will arise naturally as we design/model the individual processes.

THE PROCESS ARCHITECTURE AS SEARCHLIGHT

Pupil: You were very rude earlier about so-called process architectures that were random chunkings of organizational activity, either strung together in some sort of sequence like a kebab, or else hanging Christmas tree fashion. How is a Riva process architecture different?

Tutor: Well, we've seen how UOWs come in a variety of sizes: a *Compound* represents a much larger unit of work than an *Assay*. We can also see a partial ordering ('generates') amongst UOWs. It is likely, though not certain, that as we follow the 'generates' relationships we'll find our way to 'smaller' and 'smaller' UOWs. Look at the (real-life) example I've drawn in Figure 6.14 and tell me what you see.

Pupil: Well, I see it shows a sequence of UOWs that get smaller in 'size': I know a clinical trial is a massive thing, a CRF page is … just a page in a document called a CRF.

Tutor: Right, and all of these entities have a similarly long duration: a patient is a patient as long as the trial persists, as do the records about them. Indeed, we might reckon that the records last longer than the patient in the trial in that even though the patient has left the trial and ceased to be a patient, the records about them are kept and looked after a great deal longer.

I've drawn another sequence in Figure 6.15 to do with the administration of a university faculty. This time the UOWs are getting smaller in duration.

Can you see how this says that, to add more 'detail' to our process architecture, we must add more UOWs to our UOW diagram, with the new UOWs being generated by (probably) 'larger' UOWs already on the diagram? Note that we don't add more detail by decomposing into lower levels of a hierarchy. We add more detail by *adding more nodes to our network*. And indeed this reflects the real world. The question is only whether we choose to include them in the model or exclude them from it. By choosing to include them we are turning our searchlight on that part of the organization's activity.

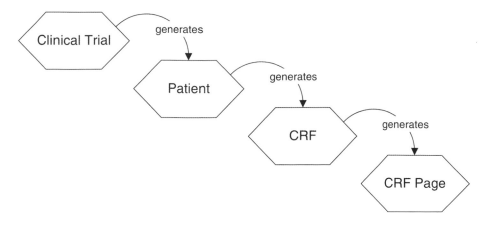

FIGURE 6.14 *Big fleas have smaller fleas …*

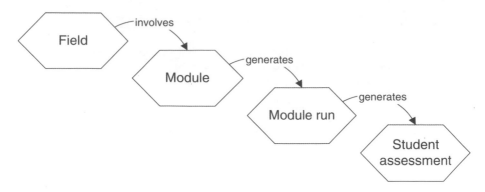

FIGURE 6.15 *Big fleas have shorter fleas . . .*

Pupil:	So, it's not that these 'smaller' UOWs are 'inside' the larger ones – more that we don't see them if we choose not to?
Tutor:	Exactly. Here's another way of looking at it. When we stare up at the night sky, we can only see the stars we can see: the ones that are bright enough. If we go somewhere where there is less light pollution, we see more of the less bright stars. It's not that we are seeing *inside* the bright stars and seeing stars there. Finally, we need a telescope to see the very faint stars between all the others – note how I said 'between'. That's how it is with UOWs and the Riva process architecture.
	Turning my night-sky metaphor on its head, I can also think of the process architecture as a searchlight which I train on the organization. I can widen or narrow its spread and look at some parts and not others. Equally I can increase or lower its intensity, allowing me to see more or fewer of the smaller (fainter) UOWs.
Pupil:	So when I choose the UOWs I choose the ones I want to choose for my purpose.
Tutor:	Right. But remember that the stars are all there – you don't invent them. You look at this part of the sky or that part. And you simply see more or fewer of them, as you choose. We're modelling reality. We're not imposing nice structures on reality.
Pupil:	You made a big deal earlier about why a sound process architecture was essential. How have we benefited from doing it the Riva way?
Tutor:	I'll give you three ways. Firstly, by deriving the architecture from the entities that are at the heart of the organization's existence we've ensured that the architecture is deduced solely from *the business the organization is in*. We haven't considered at all how the organization chooses to do business: its organizational structure, its geography, its culture, whether it is a command-and-control or empowered or consensus organization, whether it operates strict hierarchical procedures or allows cross-functional communications, whether information is treated as shared or on a need-to-know basis, and so on.
	Interestingly, we haven't been at all concerned with the *goals* of the organization either. Goals are achieved by the way that processes are done. The Riva process architecture tells us what processes we must have to do

	our business, and we use the goals when we design those processes to ensure that they do what we want.
Pupil:	In other words, the process architecture says 'If you are in this business you will have these processes in these dynamic relationships,' and it's only when we start to look inside the individual processes and at how their relationships are implemented that we introduce culture and organizational style.
Tutor:	Yes. We do that by our choice of roles and the interactions we require between them, in particular in the use of approval, delegation, reporting, agreement, authorization, negotiation, questioning and informing, and the mechanisms for these interactions.
	Secondly, by mapping the relationships between UOWs into relationships between their respective CPs and CMPs we ensure that the dynamic content of the architecture matches what is happening in the real world. Our architecture doesn't impose structure on our processes in the form of some arbitrary decomposition with no analogue in the real world.
Pupil:	And the third benefit?
Tutor:	Well, by recognising those two well-defined ways in which processes can be related – activation and interaction – we again capture reality, and moreover we do it in a way that can be directly modelled when we look at the detail within processes in a RAD.
Pupil:	The message I'm getting is that a Riva process architecture would 'survive' re-engineering, in other words it's in some sense an *invariant* of the organization; we've divided the organizational activity along the natural cleavage lines and as a result our processes will have greater cohesion and less coupling; and we've an approach that allows us to turn the spotlight wherever is appropriate to our concern.
Tutor:	Precisely. I can put it more mundanely: we can really rely on a Riva process architecture. As long as the organization stays in the same business, it will have those processes in those relationships. And if it changes its business, we can very easily determine the changes in the architecture: 'changing your business' means adding or subtracting EBEs, which – we know – will tell us immediately which new processes appear and which existing processes disappear.

CASE STUDY 2

Let's start with a relatively straightforward example of a process architecture. Suppose we are looking at the job management part of a water utility, the part that undertakes various works on its assets: drains, sewers, water supplies, reservoirs etc.

A workshop brainstormed the following as potential EBEs, and decided that the ones in brackets were either not UOWs or were not UOWs that came within the scope of the BPM project:

- Customer;
- Contact (by a customer);

- Job;
- Sample (of water);
- Inspection;
- Customer notification;
- Material requisition (MR);
- Asset;
- History of an asset;
- (Statutory notification);
- (Event/Incident);
- (Appointment);
- (Meter reading).

This led to the UOW model in Figure 6.16. Not surprisingly, given the scope of the study, the searchlight is on *Job* and its UOW 'neighbours'.

This in turn led to the process architecture in Figure 6.17.

Discussions at the workshop led to the following decisions:

- The UOW *Asset* was dropped because the processes concerned with assets were considered outside the scope of the model: they were the business of the engineers who specified, procured and maintained them.

- Various sources of requests for *Jobs* (including the periodic maintenance of assets and planned requests arising from *History*, e.g. a trend of leaks in a particular area) were bundled and shown as a cloud to represent an external source of no interest other than as a source of requests.

- All of the UOWs were supported by service functions so CMPs appeared for each.

- The relationship between **Handle a Job** and **Manage Customer notifications** is a little different from the usual. Here, in one request, **Handle a Job** requests **Manage Notifications flow** to organize the notifications for all customers. **Manage Notifications flow** is given a single geographical location, determines all the customers to whom notifications should be sent, and then sends a notification to each. It is probably not worth separating out **Manage Notifications flow** and **Handle a Customer notification**, the latter being a very 'small' process, so they have been replaced by a single process **Handle Customer notifications** which takes a geographical location and sends out all notifications for that location. This is shown in the second-cut process architecture below.

- Note also how **Handle a Job** is not interested in knowing the outcome of the notification of customers, so there is no closure of that request.

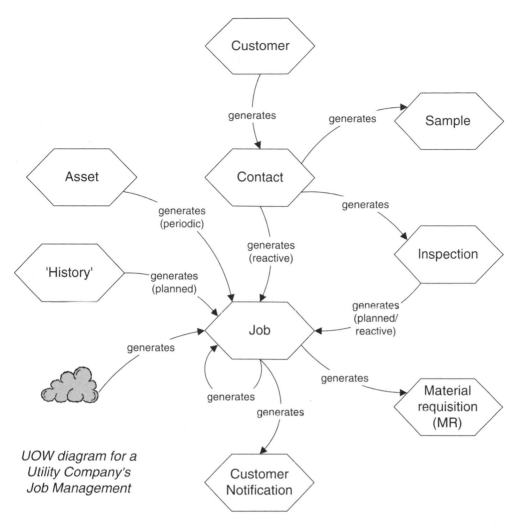

*UOW diagram for a
Utility Company's
Job Management*

FIGURE 6.16 *UOW model for a utility company's job management*

- The lifetime of customers was deemed out of scope so **Manage Customers flow** and **Handle a Customer** were dropped.
- Some reduction is possible, knowing more about the real situation. In particular, all contacts are made by telephone and it is the telephone system itself that effectively does such contact flow management as is necessary. For this reason the **Manage Contacts Flow** (case management) process can be dropped.

The result is shown in the second-cut process architecture in Figure 6.18.

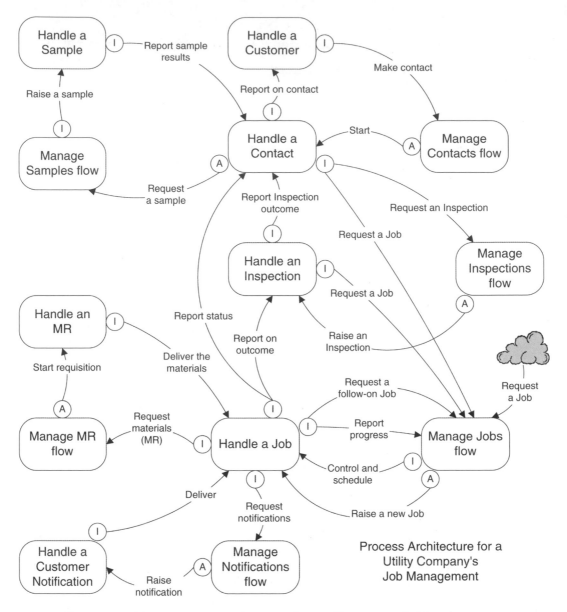

FIGURE 6.17 *First-cut process architecture for a utility company's job management*

KEY POINTS

The process architecture is a searchlight, focusing our attention on the area of the organizational activity that we are interested in.

By including 'fainter' units of work (typically designed) we increase the intensity of the searchlight.

There is no sense in which we are successively decomposing processes. We are adding more nodes to the network of processes.

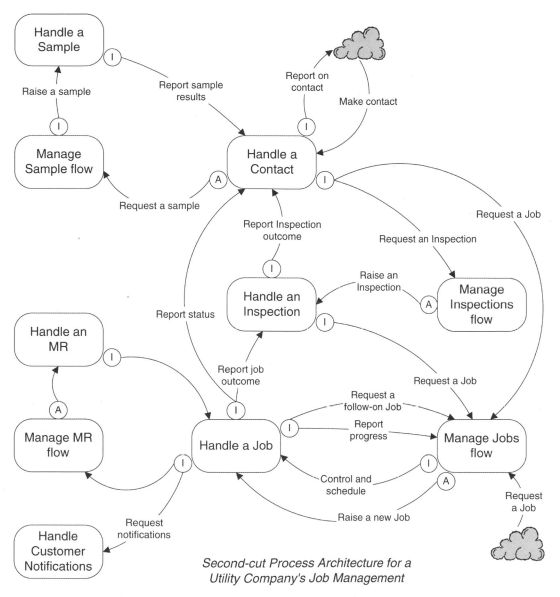

*Second-cut Process Architecture for a
Utility Company's Job Management*

FIGURE 6.18 *Second-cut process architecture for the utility company's job management*

Later, next to the water-cooler

Pupil:	I know that people get very worried about analysis paralysis. Isn't there a danger that people will perceive all this process architecture stuff as simply getting in the way of getting on and looking at the processes themselves?
Tutor:	Yes, there is. People can be very keen to get on with what they perceive to be 'real work'. I hope that I've made the case for preparing a process architecture whatever situation we are in. What I now have to do is satisfy you – and those folks with busy lives and hard targets – that preparing a process architecture can be a very quick activity.

	Remember firstly that we do it principally to make sure that we get the right chunking of all the organizational activity, and that we identify at least the dynamic relationships between the chunks – the processes.
Pupil:	OK.
Tutor:	Then remember that the work centred around brainstorming the EBEs, filtering them into UOWs, and finally doing any appropriate reduction of the first-cut architecture to the second-cut architecture. To give you a flavour of how quickly that can be done, let me give you an example.

I was at a sales meeting with a client, trying to sell them Riva-based consultancy. One of the client team had said she only had a short amount of time as she had a video-conference with colleagues across the Atlantic at which they would discuss a new framework for their QMS. She needed to get away from the meeting early in order to draft a framework for discussion. 'How long have you got before your video-conference?' I asked. 'Two hours,' came the reply. 'That's plenty for what you want: let's spend the first hour brainstorming EBEs and deriving UOWs,' I said, 'and the second hour deducing the process architecture, and reducing it as much as we can in the time.' My challenge of free consultancy was accepted, and we did it. She went off with a second-cut process architecture to get her discussions started. I don't want to suggest that two hours is all you'll ever need, but we are talking about days and not months.

CASE STUDY 3

Finally, let's examine a simple process architecture and look in particular at how the dynamic relationships would appear in RADs for them.

Suppose a software product company has a range of products. During the lifetime of a product, changes to it are proposed. Occasionally the outstanding changes proposed for a product are reviewed; some are deleted and some are incorporated in a new release of the product. A brainstorming of EBEs might come up with the following:

- Product;
- Change proposal;
- Release;
- Sale;
- Customer.

Which of these are UOWs? In other words, which of them have a lifetime which must be serviced by our 'organization'? For this exercise, we might decide that we are not concerned with selling and marketing, only the generation of product for sale. Our list of UOWs is therefore:

- Product;
- Change proposal;
- Release.

As far as the development group is concerned, ideas for new products come from outside, perhaps a marketing function within the company. During the lifetime of a product, releases are produced. Change proposals relating to products come from outside the group (customers in particular, we might suspect) and are raised during the management of the product itself. Our UOW diagram would look like Figure 6.19.

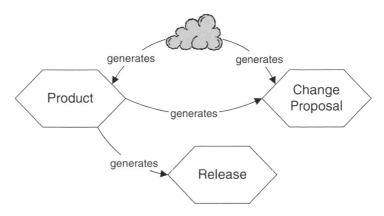

FIGURE 6.19 *UOW diagram for a development group in a software product company*

From this we can deduce our first-cut process architecture, as in Figure 6.20.

Each product manages its own releases and since releases are done one at a time for a given product, the CMP **Manage the flow of releases** doesn't exist so we can drop it. We can also assume that the business of deciding on new products – **Manage the flow of Products** – is outside the area lit by our searchlight. So we get the second-cut process architecture in Figure 6.21.

Now we can start to get inside the individual processes. Figures 6.22 through 6.25 are incomplete RADs for the four processes: we have concentrated only on the activity around the relationships between the processes – we aren't interested here in the minutiae of change management. Let's listen in on our Tutor and Pupil discussing the models.

Tutor:	**Why don't you walk us through the case study?**
Pupil:	**OK. Looking at Handle a Product (Figure 6.22), first of all I would guess that somewhere else there's a Manage the flow of Products CMP that will create the instance of Product Management Team that the RAD shows.**
	I can see that the Product Management Team has three main threads of activity. One (top right) is the main thread that gets under way immediately: the Team does some planning and then arranges for the first release; monthly it reviews things and decides whether a new release is called for, given the changes waiting their turn; and bi-annually it decides whether

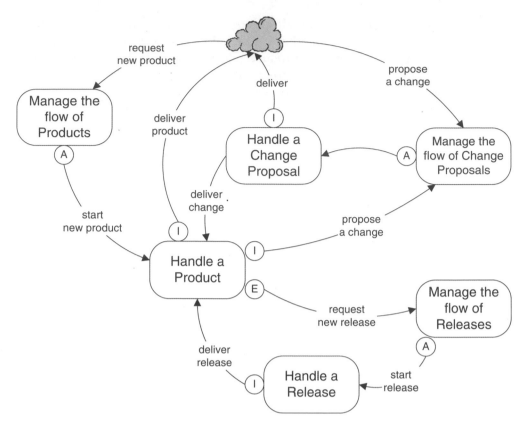

FIGURE 6.20 *First-cut process architecture from Figure 6.19*

to withdraw the product from the market or perhaps make some changes – probably quite radical ones, if any, I would guess.

If the Team decides ... we don't have any detail here ... that it wants to propose a change to the product, it uses the appropriate route like everyone else and interacts with the *Change Coordinator* ... that's the 'propose a change' interaction with the CMP *Manage the flow of Change Proposals* of course.

Going back to the point where the Team generates a new release ... we know that case management for releases is null, so the Product Management Team simply activates the CP itself by instantiating its lead role – *Release Coordinating* – it creates the responsibility for the release ... and lets it get on with it, waiting until the release has been done before checking what effect it has on the market. Interestingly, that thread – waiting to see the effect on the market – could still be in progress when the next monthly review is started for the next release?

Tutor: True. Exactly the sort of concurrency we find: things rarely happen one at a time. It's entirely possible for several releases to be in hand at one time. Let's move on to the CP for *Release*.

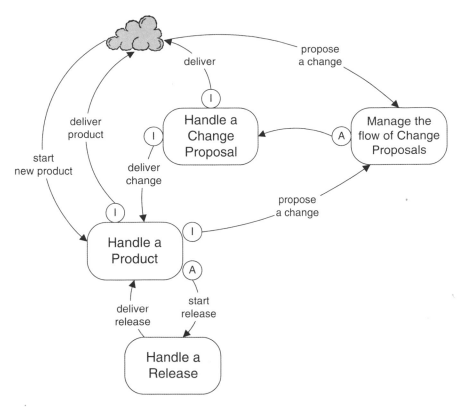

FIGURE 6.21 *Second-cut process architecture from Figure 6.20*

Pupil: In Handle a Release (Figure 6.23), I can see the instantiated *Release Coordinating* role preparing the release by filtering the outstanding change proposals, getting the appropriate software changes made for all of the change proposals that are accepted for the release, postponing some proposals for a future release presumably, and rejecting others there and then. Presumably that rejection is picked up in Handle a Change Proposal ... I can check that process interaction in a minute. Once the release has been made there's an interesting state – *New release ready* – which of course the Product Management Team are waiting to hear about back in Handle a Product in Figure 6.22. And there's an interesting little ... trick? We need to signal that each individual change that made it into the release has indeed made it, presumably so that the Change Coordinator can take the appropriate action over in the appropriate instance of the CP Handle a Change.

Tutor: Well, you might call that a trick but it captures reality!

Pupil: I noticed that the process architecture shows the new release being 'delivered' to the product via an interaction between Handle a Release and Handle a Product. We've actually modelled this in your minimalist way as a state–trigger pair: the state *New release ready* in Handle a Release is spotted as an external event in Handle a Product. I guess we could have chosen to be more explicit and modelled the interaction that – say – takes

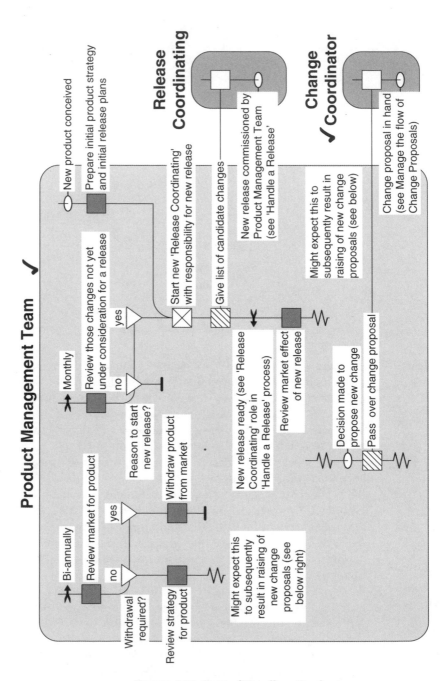

FIGURE 6.22 *Part of* **Handle a Product**

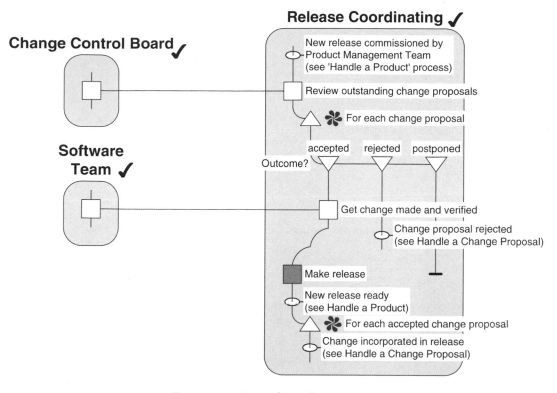

FIGURE 6.23 *Part of* **Handle a Release**

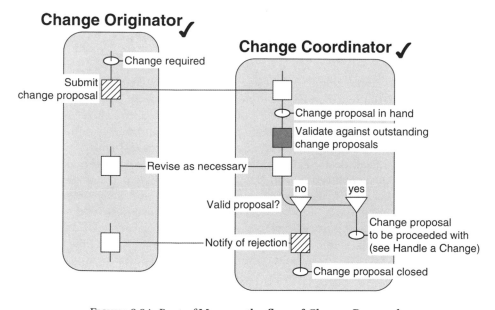

FIGURE 6.24 *Part of* **Manage the flow of Change Proposals**

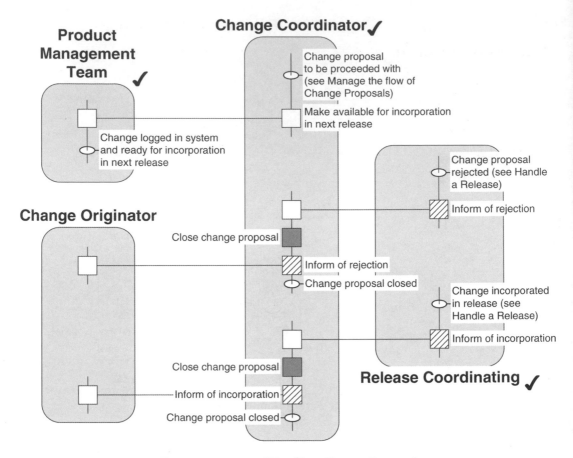

FIGURE 6.25 *Part of* **Handle a Change Proposal**

Tutor:
Pupil:

place between the Release Coordinator and the Product Management Team to notify delivery?

Yes, we could – that's a modelling decision of course.

Another thing I noticed was that the responsibility for a new release has been represented as an abstract role – *Release coordinating* – and that the Product Management Team instantiates it for each release. But the same isn't true for a change proposal: we're not instantiating a role – a responsibility – just for that change proposal. Instead, we're passing it to a fixed post, the Change Coordinator ... which I feel makes sense, but I'm not convinced ... wait ... most of the handling of a change proposal actually goes on in the handling of a release. The case management of change proposals is little more than a matter of checking that the request is a valid one and then sticking it on the pile of change proposals waiting for a release ... oh, and notifying people of the outcome. The Change Coordinator does the case management and deals with individual cases – which is, I guess, why it makes sense not to have an abstract role taking responsibility for an individual change proposal.

Tutor:	Yes, that's fair – in designing these processes we've chosen to map the responsibility for all change proposals and for their case management onto a single post in the organization. In a different situation we might have modelled it differently. With the model as it stands we haven't yet said how the responsibility for a release is going to be allocated. That's a design decision we still have to make.
Tutor:	Let's move on to the CP and CMP for Change Proposals.
Pupil:	In both Manage the flow of Change Proposals (Figure 6.24) and Handle a Change Proposal (Figure 6.25), we've got a role called *Change Originator*. I guess this is a sort of anonymous role, in that a change can be originated by a Product Management Team, for instance, or by any agent in 'the outside world'.
Tutor:	That's right – remember the cloud in the second-cut process architecture: proposals for changes can come from there. So, let's start with the CP Handle a Change Proposal (Figure 6.25).
Pupil:	The top thread is where the CMP has started a new case ... and all the Change Coordinator does is pass the proposal to the Product Management Team, who put it on the pile ready for consideration at the next monthly review. That's an information-passing interaction between Handle a Change Proposal and Handle a Product.
	Then there are two process interactions with Handle a Release where we hear about the rejection (*Change proposal rejected*) or incorporation (*Change incorporated in release*) of a Change Proposal. In both situations, the case ... the Change Proposal ... is closed and the outcome is notified to the original Change Requester. I suppose both of these process interactions are the result of design decisions about how we want these processes to operate, so they weren't on the process architecture?
Tutor:	That's right. We must always remember that the purpose of the process architecture is to give us our initial chunking of the organizational activity into processes and also to tell us the dynamic relationships that must be there. When we decide how we want the processes to operate we might well generate the need for additional process interactions, very often for information-passing.
Pupil:	You've been careful to separate the CP and CMP for *Change Proposal* but since, as you observed, the former does run straight through into the latter, is there really any sense in keeping the RADs separate?
Tutor:	We could indeed combine those RADs, as a modelling convenience. I've separated them, as you say, just to emphasize that there is case management going on here, and it might result in a change proposal being rejected before it is put into the pot, so to speak. In a realistic situation we could imagine far more complex decision-making being required at this stage, before a change proposal ever finds its way into the list. But certainly we could combine processes onto one RAD if it helps understanding, without losing precision. You might like to try the exercise of doing precisely that and putting all four of the processes we have just looked at onto one RAD. Be careful with pre-existing instances!

7 Dynamism in the world

Shows how the process architecture captures all the between-process concurrency in the world.

INTRODUCTION

In Chapter 1, our Tutor stressed that things happen in organizations because a mass of concurrent activity takes place, and that, thanks to instantiation, there is a flux of instances that make that concurrency happen. In Chapter 3 we looked at how a RAD captures the concurrency possible in a single process. Now that we also have an understanding of the process architecture of the organization – in particular the way that processes activate each other and interact – we can extend our horizons and look at how concurrency occurs across the world that we are concerned with: the 'organization'. This is the right moment to summarize where we capture concurrency in Riva models:

- An organization has a number of process types, each of which can be instantiated.
- Each process instance operates independently. Process instances can interact.
- Within a process instance each role type in the process can be instantiated.
- Each role instance operates independently. Role instances can interact.
- Within a role instance separate threads can be instantiated.
- Each thread instance operates independently. Thread instances can combine and divide.

I have been strict here: a process instance only exists inasmuch as we have instantiated the lead role of the process, and that role instance has in turn instantiated other roles. Similarly, we don't instantiate threads as such, rather the actions and interactions on those threads. Put another way, if we look inside a 'running' Riva model of an organization we shall see a flux of role instances, each role instance having a number of action and interaction instances in progress. This reflects precisely what is happening inside the building when we walk in: the world is a mass of acting and interacting responsibilities.

In this chapter, we shall explore the notion of concurrency through two rather different case studies. In the first – a fairly conventional situation – we examine how we can represent concurrency in a single RAD of the organizational activity, or more usefully as a set of processes deduced from the process architecture. In the second, we look at a completely different sort of process with a rather different dynamic: email.

CASE STUDY 4

In writing a book such as this, it is hard to decide at what level to pitch ·worked examples. A real example, from the real world, will be too big. A sufficiently reduced example from the real world will probably be too simple to be convincing. And the real world rarely offers tidy case studies for textbooks. I have chosen a real project that starts from exactly the sort of ill-formed input that typifies real projects, and it allows me to show the revelatory power of the Riva method.

The problem – as originally stated – was to prepare a description of 'the process' (singular) outlined in the following piece of descriptive text. I shall do what I always do: start with a process architecture, in order to chunk the organization's activity into processes. Once we have the process architecture in place we can start to look at the individual processes in it. Yes, there is more than one.

When we have finished dissecting the processes we shall model the problem as if it were indeed one process, conflating all the processes onto a single RAD.

The sentinel case study

This case study concerns the work of a European project in the control of antibiotic-resistant infections in children. Often children are asymptomatic carriers of bacteria that they can pass on to other children by playing with the same toys, touching each other etc.

The medical infrastructure in place to monitor, contain and destroy the infectious organism typically involves a medical practitioner who collects samples from human individuals and sends them for further analysis. Consequently three levels of operation can be recognized:

1. The field level – where the medical practitioner interacts directly with the potential human hosts, collects samples and basic characterization of the individual and sends the samples for further analysis at the next level, the microbiological/biochemical laboratory.

2. The laboratory level – which corresponds to the facilities that process the samples sent in from the field level.

3. A central location – where data integration and decision-making take place.

Typically, the human host is mostly characterized at the field level and the infectious organism is characterized mostly at the laboratory level. In our case study, Care Centres, the Microbiology Centre and the Epidemiological Information Centre represent the three levels.

At the *Care Centre* (for example a nursery school), within a Care Centre Period, swabs are taken from children's noses for analysis. The Care Centre Operative collects these swabs, referred to as C Samples, together into a C Sample Batch to send to the Microbiology Centre. This is usually done one class or group of children at a time.

These are called Care Centre Units and the period when their swabs are taken is called the Care Centre Unit Period – this must fall within the Care Centre Period. The Care Centre Operative records all the necessary details (ID numbers etc) in order to ensure that the link between the Attendee (i.e. a child) and their C Sample is maintained.

At the *Microbiology Centre* (usually a laboratory) the Microbiology Centre Operative receives a batch of samples, i.e. a C Sample Batch, records them as M Sample Batches in the local system, and then reorganizes them into one or more Sample Analysis Groups to suit the process at the Microbiology Centre. (To make this point clearer imagine that one Care Centre Unit sends 19 swabs to the Microbiology Centre, another sends 22 and a third 26. The Microbiology Centre can process, say, 35 swabs at a time. The three distinct M Sample Batches – of 19, 22 and 26 – are reorganized into two Sample Analysis Groups of 32 and 35 respectively.) Each swab is cultured on a Petri dish and an attempt is made to isolate different strains of bacteria and then to identify them. This is done using various tests, the main ones being by serotype, by antibiogram, and by pulse field gel electrophoresis (PFGE). Information about strains identified in each swab is sent back to the Care Centre.

Information about strains is also sent to the *Epidemiological Information Centre.* The role of this centre is partly to organize and manage the overall campaigns (e.g. to define the Care Centre Periods) but, more importantly, it determines whether alarms must be raised and to what level. For example, if any antibiotic resistant strains are present, an alarm is raised. However, actions at the Care Centre following an alarm vary from country to country. In Sweden the children must be sent home and the parents are paid by the state to stay at home with them. In Portugal they are also likely to be sent home, but they return earlier because there is no social provision. If a situation is deemed serious enough, a Care Centre may be closed for a time.

Analysis

Let's start by listing candidate EBEs using only the text provided. Clearly in a normal process study we would be running brainstorming workshops. When faced with this sort of situation, spotting nouns is a productive approach:

- Child;
- Medical Practitioner;
- Sample;
- Laboratory;
- Care Centre;
- Microbiology Centre;
- Epidemiological Information Centre;
- Care Centre Period;
- Care Centre Operative;
- C Sample;
- C Sample Batch;
- M Sample Batch;
- Class (of children);
- Care Centre Unit;
- Care Centre Unit Period;
- Attendee;
- Sample Analysis Group;
- Swab;
- Bacterium;
- Strain of bacteria;
- Petri dish;
- Test (e.g. PFGE);
- C Sample Result;
- Campaign;
- Alarm;
- Resistant strain;
- Country.

You can see I have already been careful to ensure they are all entities: we can meaningfully put the word 'a' or 'the' in front of each.

The next step is to filter them using the guidelines listed in Chapter 6 to get to a list of UOWs for this 'organization'. Because we want to model the organization as-is, we include designed as well as essential UOWs. Bulleted items are those that passed through the filters. Bracketed items didn't and the reasons appear against them. (This list looks very much the way it did on the whiteboard during the EBE brainstorming session.)

> (Child) = C Sample: we are interested in the life history of the sample the child gives but not of the child. If anything, the child is only a role that plays a part.

(Medical practitioner): a role that plays a part, lifetime not of interest to us.

(Sample): see C Sample.

(Laboratory): lifetime not of interest to us.

(Microbiology Centre): lifetime not of interest to us.

(Epidemiological Information Centre): lifetime not of interest to us.

(Care Centre): lifetime not of interest to us.

(Care Centre Period): a period of time.

(Care Centre Operative): a role that plays a part, lifetime not of interest to us.

- C Sample.
- C Sample Batch.

  (M Sample Batch) = C Sample Batch.

  (Class (of children)): lifetime not of interest to us.
- Care Centre Unit.

  (Care Centre Unit Period): a period of time.

  (Attendee): see Child.
- Sample Analysis Group.

  (Swab) = C Sample.

  (Bacterium): lifetime not of interest to us.

  (Strain of Bacteria): lifetime not of interest to us.

  (Petri Dish): lifetime not of interest to us.

  (Test): lifetime not of interest to us.
- C Sample Result.
- Campaign.

  (Alarm) lifetime not of interest to us.

  (Resistant Strain) lifetime not of interest to us.

  (Country) lifetime not of interest to us.

During the break, next to the water cooler

Pupil: I was concerned when you so easily knocked *Child* out of the list of UOWs – surely this whole thing is about children?

Tutor: Well, is it? Remember that when we decide on the EBEs and filter out the UOWs, we choose those that characterize the organization we are interested in – we turn the searchlight on some areas and not on others. We also adjust the intensity of the searchlight, perhaps differently in different areas, to illuminate just those things that are 'large enough' for our purposes. You might feel that my choice in the filtering would not be your choice: that simply means we are thinking about different 'organizations'. We could

	widen our interest a bit and bring in how we deal with children who are found to be carriers for instance.
Pupil:	OK, so you're saying that that's a scoping decision. If we decide we want to include that part of the business, we shall indeed have *Child* in the list and ... there will be a process called Handle a child and another called Manage the flow of children (though I'm not too sure what that would be about in this situation).
Tutor:	Exactly. We could then cover what happens to each child, in particular if they are found to be a carrier or an alarm is raised. But you might agree that the description we have been given doesn't get into any of that, so we should scope it out by leaving *Child* out of the list of UOWs.
Pupil:	OK. Then following the same reasoning, *Alarm* was an interesting candidate. Alarms have a lifetime of interest to someone ... OK, but not us, we're saying. They are the business of national bodies. I guess I've just made a decision about where the searchlight is pointing and how wide I want its beam to be.
	I also thought *Sample Analysis Group* was interesting. Sample Analysis Groups are – as the text makes clear – only there to suit the way the Microbiology Centre works: they are designed EBEs.
Tutor:	Right. We're faced with the typical decision: are we concerned with developing a pure process architecture which is independent of how we choose to do things, or do we want to include designed mechanisms? In this instance, our concern is with the way things are. If we were looking for re-engineering opportunities we might take the opposite view. In later sessions, we shall look at how our purpose will determine which way decisions like this should go.

We can now transfer the six surviving UOWs to a UOW diagram, adding the 'generates' relationships that hold between them. The result is Figure 7.1.

Let's unpick what this is saying:

- The 'big' unit of work is the Campaign. Campaigns are defined at the Epidemiological Information Centre.

- During each Campaign a number of Care Centre Units are generated. This is done within the handling of the Campaign – no outside service is involved, so the relationship is a task force relationship.

- A Care Centre Unit generates C Sample Batches. The latter are the units in which C Samples are sent to the Microbiology Centre, which deals with these as a service. The Microbiology Centre has them coming in from many Care Centres.

- C/M Sample Batches are restructured as Sample Analysis Groups for the 'convenience' of the Microbiology Centre. Note the $n:m$ cardinality. Sample Analysis Groups are dealt with as a service.

- C Samples arise from the Care Centre's own handling of C Sample Batches. There is no separate service dealing with C Samples for them and others. This is a task force relationship.

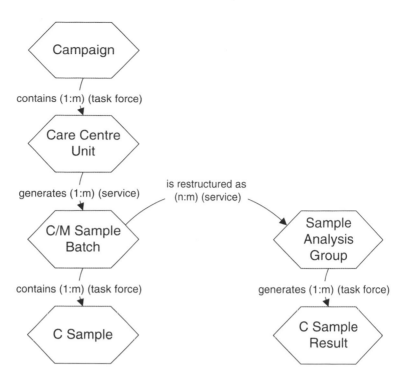

FIGURE 7.1 *UOW diagram for the Sentinel case study*

- During the lifetime of a Sample Analysis Group, the results for the constituent C Samples are generated. C Sample Results are managed within the context of each Sample Analysis group so this is also a task force relationship.

We might first think that C Sample Results arise from (are generated by) C Samples. We could certainly say that C Sample Results are 'related to' their corresponding C Samples, in the obvious way. But that is not the dynamic 'generates' relationship that we are looking for, and which does exist between Sample Analysis Group and C Sample Result: it is during the lifetime of the Sample Analysis Group, which is processed in Microbiology, that the C Sample Results are generated.

Note how in this UOW diagram things get smaller as we follow the arrows.

The next step is to mechanically transform the UOW diagram into the first-cut process architecture in Figure 7.2.

Let's walk through this. (Note that we have not added the CSPs. This is common but not inevitable. The CSPs sit 'around' their CPs and CMPs, and this study was concerned with operational activity, rather than strategic planning.)

There is a 'root process' – **Manage the flow of Campaigns** – which, at appropriate moments, activates **Handle a Campaign** in order to get a

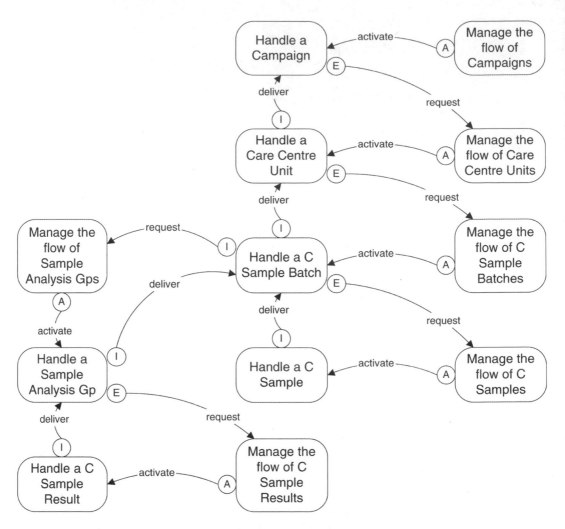

FIGURE 7.2 *First-cut process architecture for the Sentinel study*

campaign done. A Campaign is made up of a number of Care Centre Units, which are managed by the campaign. The diagram therefore shows **Manage the flow of Care Centre Units** encapsulated in **Handle a Campaign. Manage the flow of Care Centre Units** has its own existence and needs to be thought about in its own right, but we are recognizing that when we come to model it, it will be 'found in' or best modelled in **Handle a Campaign** since that is where the flow management actually takes place.

During **Handle a Care Centre Unit**, C Sample Batches are generated in task force mode. C/M Sample Batches are then converted at the Microbiology Centre into Sample Analysis Groups for which there is separate flow management. **Handle a C Sample Batch** makes its request to **Manage the flow of Sample Analysis Groups** which activates **Handle a Sample Analysis Group** when ready, which in turn goes to **Manage the**

flow of C Sample Results for each C Sample Result it requires. There is a chain of deliveries back through the sequence.

We can now fold processes together where a task force relationship is involved. This yields the second-cut process architecture in Figure 7.3.

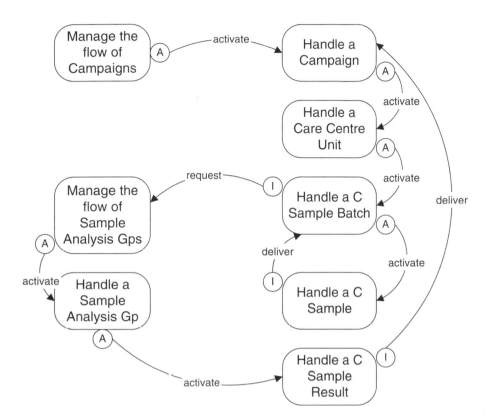

FIGURE 7.3 *A second-cut process architecture for the Sentinel study*

What has happened?

- We have folded each encapsulated CMP into the single CP that uses it.
- We have recognized that the chain of delivery from **Handle a C Sample Result** to **Handle a C Sample Batch** to **Handle a Care Centre Unit** to **Handle a Campaign** can be short-circuited, leaving one 'delivers' relationship from **Handle a C Sample Result** to **Handle a Campaign**. Remember that the first-cut architecture is derived mechanically from the UOW diagram and needs to be put against the world. It is **Handle a Campaign** that wants the results in order to determine whether an alarm needs to be raised.

The net result is that we can recognize eight discrete processes that are covered in the one description. It is certainly possible to draw a RAD as if this piece of the world really were a single process – Figure 7.4 does precisely this, in one way. Spend a few moments chunking the RAD along

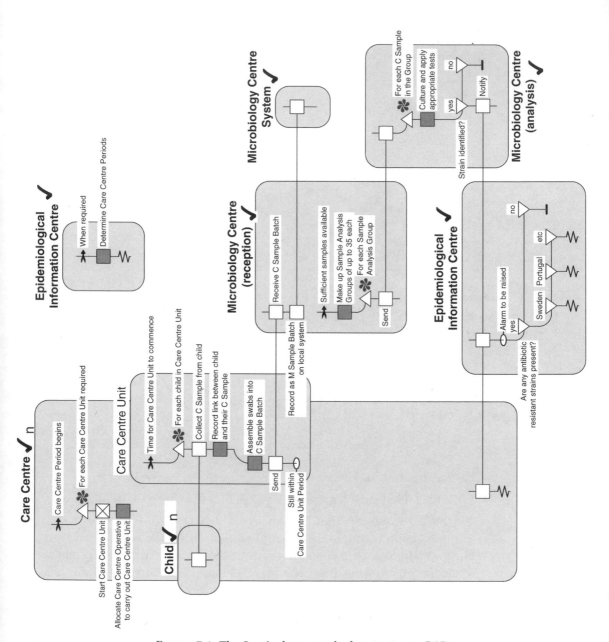

FIGURE 7.4 *The Sentinel case study drawn on one RAD*

the lines of Figure 7.3. You will find all eight processes in there – some of them consisting of a single interaction. You will also be able to see the activation and interaction of the processes: sometimes as a role instantiation and sometimes as an interaction between roles. The concurrency that builds up in the process when things get going is created from the instantiation of Care Centre Units and the multiple instantiation of new threads in various roles thanks to replicated part refinements.

Later, next to the water cooler

Tutor: I hope you're beginning to see how, by taking the process architecture view of things, we get a much more realistic understanding of the mass of concurrent activity that the organization indulges in.

Pupil: Yes, I can. It was interesting to see how the RAD you drew looks rather sequential at first glance but does in fact capture how – when this part of the world 'runs' – a mass of concurrent activity spreads out and then dies away. I can see that it's the replicated part refinements that do that: they create lots of concurrent threads of activity. But the process architecture view and the eight separate processes really make that so much clearer. I can visualize how a single Campaign starts, kicks off several Care Centre Units, each of which ... etc etc, and before we know it there are scores of process instances running independently. In time they die off and the world is quiet again.

Tutor: Yes, people all too easily look at an area of organizational activity and draw it as a sequential thread as if that is what happens. If they think a little harder they might perhaps use a single RAD to capture some or all of the concurrency that is actually going on. This is better – we are getting closer to reality. But if they start with a process architecture then they are thinking 'concurrency' right from the beginning: the separate processes, the dynamic relationships between them that cause new instances to be started ... we're capturing concurrency before getting anywhere near the workings of roles and so on. When we start to open up the individual processes we can explore the concurrency between roles, and by opening up roles we can explore the concurrency within a role. So by taking concurrency in three steps – architecture first, then processes, then roles – we get a firmer and more accurate handle on it.

I once saw a flowchart someone had drawn of their 'process' – in essence it was the sequence of boxes shown in Figure 7.5. Would you care to criticize it with your new understanding?

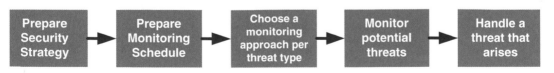

FIGURE 7.5 *Not what happens in the real world*

Pupil: Well, the first thing is that I'm sure that they didn't do things in the strict order shown: devise a Security Strategy, then devise a Monitoring Schedule, then choose a way of monitoring one sort of threat, then spot a threat, then do something about it. Does the one security Strategy last forever? Does the one Monitoring Schedule last forever? Does monitoring for threats cease once the first threat is spotted? What happens when the second threat comes along? The more I think about it, the more unrealistic it becomes.

Tutor: You're right; it's not what they did, of course. Moreover, in the small print it was clear that the Monitoring Schedule was maintained and adjusted over time. It had its own independent life. What about UOWs?

Pupil: Just what I was thinking. A Threat is a UOW, to name one. So there will be a CP for that – Handle a threat – and a CMP – Manage the flow of threats. The Monitoring Schedule sounds like something with its own life and since there is only one, there won't be any case management. Ditto for the Security Strategy. So we probably have four interacting processes here instead of one strange sequential one.

OK, I really ought to start with a UOW diagram. Let me sketch one in Figure 7.6 just from the bare facts that we have.

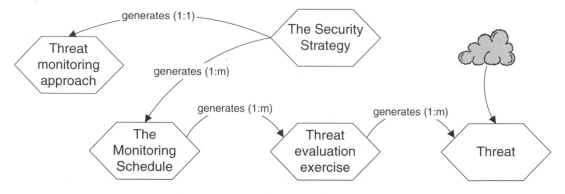

FIGURE 7.6 *A UOW diagram for the 'process'*

I've guessed that the Monitoring Schedule lists Threat evaluation exercises which take place and reveal threats. I guess threats will also appear out of the blue. So, after a bit of work the process architecture will probably look like Figure 7.7.

I haven't added subsidiary interactions but I guess it's enough to demonstrate that the flowchart process really didn't capture reality at all.

Tutor: Exactly. Let's pull this together.

If we string together things that do occasionally happen in a particular order, we shall not capture reality; in particular we shall probably miss the potential concurrency.

If we mix up UOWs, we shall not capture reality. It will be plain wrong.

If we conflate CPs and CMPs, we shall not capture reality. We shall miss concurrency and most probably get it just plain wrong!

If we don't start with a UOW analysis and derive the process architecture

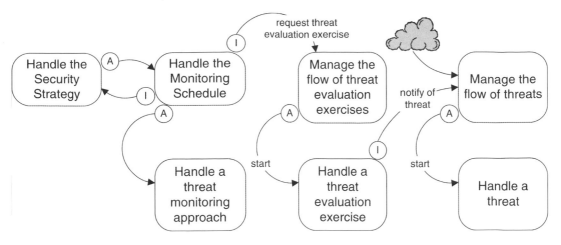

FIGURE 7.7 *A process architecture for the 'process'*

from that, we shall end up ignoring the potentially concurrent processes and describing a world that does not exist. The resulting process model will not reflect reality and it will be impossible to use. It will not be a basis for process improvement and would be a disastrous starting point for developing systems to support the activity.

An organization really is a portfolio of processes. We can be more specific: an organization is a portfolio of end-to-end CPs together with the necessary CMPs and CSPs. That's why a Riva process architecture fully captures the organization in process terms.

DYNAMISM AT WORK

Pupil: That last case study was about … shall I say … a conventional process. But my thoughts are straying to something a little less tangible: the world of email conversations. The 'organization' here is certainly not some functional group. It's the rather abstract world of structured communication by email. I have to ask you whether you think Riva's concepts work here.

Tutor: And I have to say 'Of course!' In fact email is a very nice example of the sort of very fluid and very dynamic process that goes on in organizations and which traditional flowcharts and other notations simply can't deal with. Let's give it a go. Informally, I could describe the world of email as one in which people can start conversations. A 'conversation' is just an area that someone wants to start some 'threads' going on.

Suppose I'm interested in having a conversation about William Morris and the Arts and Crafts movement. I might start three threads of conversation: one on his Red House pattern designs, one on his Kelmscott books, and another on his Socialist writings. I start each thread simply by sending out an email to some people I think might be interested in it. Each recipient picks up the thread either by responding to it – perhaps to the same group of

Pupil: people or a different group, by ignoring it, or perhaps by sending out an email to another group of recipients on a new thread of their own within the conversation. That sounds like a nice dynamic, fluid sort of process. Where shall we start in our analysis?

Well, as good Riva analysts, our first instinct will be to draw a UOW diagram (Figure 7.8). I'll assume that an eConversation can start spontaneously, in

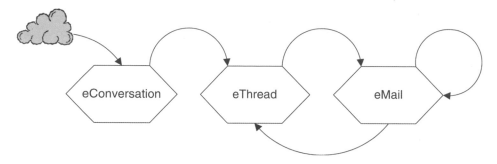

FIGURE 7.8 *UOW diagram for an email system*

someone's mind. And that an eConversation starts a new eThread with a particular message title. Let's equate each eThread with a message title. Then each eConversation evolves in the form of one or more eThreads, where an eThread takes the form of one or more eMails. Obviously, eMails can generate new eMails and/or eThreads.

Our second instinct will be to transform this into a first-cut architecture (Figure 7.9).

Tutor: Good. Now reduce it to the second-cut architecture.

Pupil: I can see some obvious reductions: 'delivery' resulting from 'generates' means nothing in this context, so we can remove those process interactions. All case management is clearly done in task force mode and so a CP does the CMP for generated UOWs. I guess that all case management is null anyway in this simple case ... but I can imagine useful facilities to do with logging and measurement and so on. Anyway, we seem to be just left with a set of 'activates' relationships between the three case processes reflecting the UOW structure (Figure 7.10).

Should we be surprised by the final, second-cut architecture?

Tutor: Not in the slightest. Because there is no case management, the process architecture will have exactly the same structure as the UOW diagram. The real point that we can draw from this example is that this simple picture represents completely the way that things can blossom – even explode – on email. A conversation can develop in an infinity of ways. One conversation might consist of a single thread of a single message that causes no further email. Others can cause torrents of new threads that all have their origin in that first spontaneous email. One day, perhaps, all the threads peter out and the conversation as a whole 'ends' at that moment. This picture captures the cascade of process instances that really does happen in the real world. Email is not a serial affair and it's not workflow.

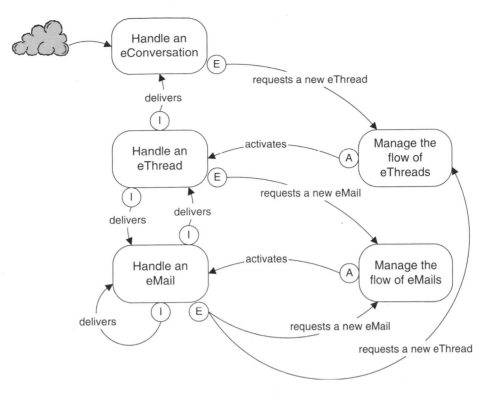

FIGURE 7.9 *First-cut process architecture for an email system*

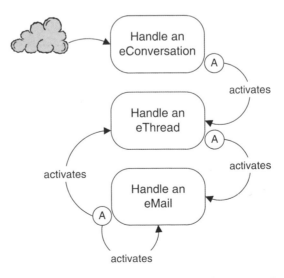

FIGURE 7.10 *Second-cut process architecture for an email system*

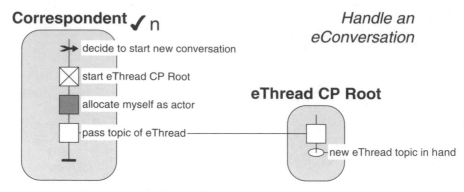

FIGURE 7.11 *The Handle an eConversation case process*

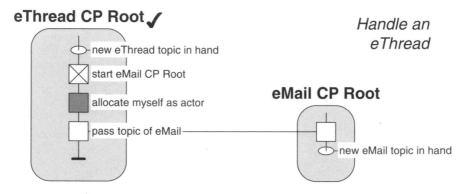

FIGURE 7.12 *The Handle an eThread case process*

	But why don't you take it one step further and draw some RADs for the three CPs that we're left with?
Pupil:	OK. Three CPs. (Figures 7.11 to 7.13).
Tutor:	Now we can start to see how a RAD captures all the possible future behaviours of the 'organization'. Talk me through what you've drawn.
Pupil:	Well a conversation starts spontaneously with a Correspondent who creates a new 'responsibility' for a thread and takes on that responsibility (role instance) themselves. That role instance then creates yet another new responsibility in the form of an instance of a role for an email, and again acts that instance.
Tutor:	OK, that's the first two CPs. Then what?
Pupil:	In that responsibility (eMail CP Root) the correspondent then actually writes the email, chooses a set of recipients from an Address Book, and sends each of them a copy. The interesting thing is that we have identified being the recipient of an email with the responsibility for dealing with it; though ignoring it is one way they can deal with it, as ever! And these (very tiny) responsibilities are created dynamically.

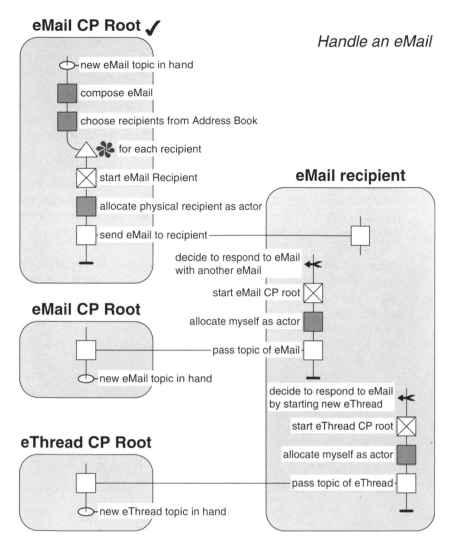

FIGURE 7.13 *The Handle an eMail case process*

Tutor: Exactly. Then what?

Pupil: The recipient doesn't actually have to do anything – they can ignore the email. But they can also decide to contribute to the thread by sending out a response email – again creating new responsibilities (role instances) for each recipient – or they can decide to start a whole new thread in the conversation.

Tutor: Right. In fact, of course, the RAD says that they can contribute to the existing thread *and* they can start a new thread ... and indeed they can do those two things as many times as they like – just as in real life: we leave a message in our inbox and two days later decide to send out a reply, and then the next day decide to forward it with a different reply to a different group of people, and so on. The conversation can 'blossom' in the hands of recipients in an infinite number of ways. How does the concurrency show itself?

Pupil: I guess it's reflected in the multiplication of instances of the roles for conversations, threads and emails. Over time, these are generated by people sending emails and creating them. Once a conversation starts flowing you could have lots and lots of these at any one moment, all operating independently and deciding how they want to push things on, if at all.

Tutor: Yes. Now even though our process architecture is 'small', it does highlight concurrency in terms of the separate *processes* that can be instantiated. There is, of course, nothing to stop me then drawing a single RAD combining all three processes – this is a modelling choice, remember, and not a statement about the world. If we do that, we'll get something like Figure 7.14. It describes that same behaviour as Figures 7.11 to 7.13 but we lose the visibility of the separate processes, and we might regard that as a bad modelling choice as a result.

One thing that is currently fixed in the process we have drawn is the number of pre-existing instances of *Correspondent*; in other words we start with – and remain with – a fixed number of correspondents. This means we're actually missing a degree of variability that is present in the world. Now something that we haven't looked at is the Address Book.

Pupil: Well, we've – I've – assumed that it's just there and available to all. Isn't that fair?

Tutor: Yes, we could certainly regard it as some sort of shared resource. But – in the world of email – we can treat it as a designed UOW: it has a lifetime, it undergoes change during that lifetime, and it needs handling. And if it's a UOW then it has a CP. Indeed, we can associate the Address Book with its CP. Then if someone wants email addresses they can interact with the Handle the Address Book process.

Pupil: I like it. Presumably if we wanted to go further we could identify a 'smaller' UOW called *Account Holder* or some such, which is generated by the *Address Book* unit of work.

Tutor: Keep going ...

Pupil: Well, in our first-cut process architecture we know that there will be a Handle an Account Holder process and a Manage the flow of Account Holders process. We can associate an instance of Handle an Account Holder with an account holder. I'm really getting the feeling that the world is full of processes rather than data or objects. Any significant object – in particular any with a lifetime in which things happen – can be associated with its CP instance.

Tutor: I think this demonstrates the power of the Riva process architecture. It moves us away from thinking about data and objects and towards processes.

In true BPM style, we start to identify things with their processes. An Address Book, or an Account Holder, becomes a process rather than an object about which we keep data, or an object which processes messages. The object-oriented paradigm converted us all from thinking about data on which computation is performed, to thinking about objects which have

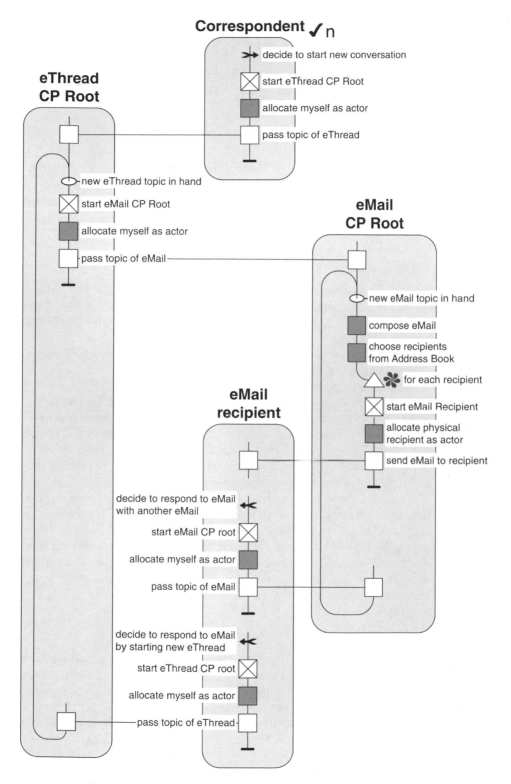

FIGURE 7.14 *The three email case processes drawn on one RAD*

associated computation (to process messages) and state. The process-oriented paradigm finally completes the inversion: everything is a process and instances have state which we can treat as data. Riva moreover adds business-related concepts and dynamics to basic object orientation – role and interaction and prop/resource in particular.

ALL THE WORLD'S A THEATRE

From the outset, there has been a theatrical theme running through the book: in particular, the notions of roles and their actors. Let's round off the metaphor and see what it tells us about the world and how Riva helps us understand it and get our arms round it. We'll leave it to our Tutor and Pupil to explore the idea.

Later, next to the water cooler

Tutor: We've been talking about activity in a part of the world, the part inhabited by our 'organization'. We can think of that part of the world as the 'theatre'. This theatre will be a rather twenty-first-century one with some unusual characteristics. Let's start with processes.

Pupil: Well, we've got processes – plays, I guess. Plays are written down in scripts ... process models and ...

Tutor: Stop there for a moment. Think about instances ...

Pupil: Plays are performed ... they have performances ... process instances. So when we walk into the theatre we'll find performances of plays going on. Ah – the theatre seems to be a multiplex, because I have lots of plays being performed at the same time! Worse, some performances start new performances! And presumably they have to find a stage to operate on.

Tutor: Concentrate on one performance for a moment.

Pupil: Well, a play consists of a number of roles. Each role ... ah, each role instance ... is acted by an actor – who was cast in that part somewhere along the line. Things start getting a bit weird around here, I suspect, because during the performance some actors might act several roles, rushing around the stage from one to another. Worse, I suppose they might be acting roles in more than one performance so they'll have to run from stage to stage, changing costume as they go. In the worst case they'll be acting several roles in several performances.

Tutor: When we're talking about performances, remember we need to talk about role instances.

Pupil: Hmm. That's a bit weird too: how many Hamlets can you have in one performance of Hamlet? Only one, I guess; but we have seen plenty of processes in which some roles are instantiated many times – mercifully, Shakespeare was content with a single instance of the Prince of Denmark on stage at any one moment. But it appears that in some plays new role instances are created while the play is in progress. And an actor has to be found to play the role instance. In extreme cases, the actors might be writing bits of play, inventing new roles, instantiating them, and then casting actors as they go along. This is truly contemporary!

Tutor: Absolutely. You mention casting ...

Pupil: Yeeees. We said, didn't we, that the allocation of actors to role instances is just more process ... so casting happens on stage and possibly during the performance!

Tutor: Any thoughts about props?

Pupil: Well ... the props are the resources actors need to play the role instance. It might be a book or a newspaper in a real play, and an information system or a software application in a real business process. The costume sounds a bit like the mindset they need for the role.

Tutor: So we have a number of stages, each with a performance of a play going on. Performances are starting up and stopping all over the place. On each stage, role instances are being played by actors, who are possibly rushing from play to play and from role instance to role instance, putting costumes on and taking costumes off, and picking up and putting down props as appropriate. So far so good. What about role actions and interactions?

Pupil: I'm not sure I want to watch one of these plays. Sometimes a role instance is doing an action: so the actor is giving a soliloquy; sometimes it's interacting with another role instance, or even with several others at the same time. There may be several soliloquies and several conversations all going on at the same time. And occasionally proceedings will get held up for want of an actor ... who might be acting another role instance on another stage. It's madness.

Tutor: Hold that thought. You've only dealt with role interactions in a single performance, but we have seen how process instances interact. And in the theatre?

Pupil: Oh dear. Some plays are connected. A performance of one play has to interact with a performance of another. That means that role instances in the two performances have to interact – there must be communication systems between the stages – telephones or email or something. Or perhaps the actors rush to and from each other's stage, or perhaps they meet in the corridor? My head is starting to spin.

Tutor: We've just described a chaotic-sounding theatre – but it's no more complex than the average organization. The problem is of course, that when most people look at an organization and are faced with the task of capturing what goes on, they find it so overwhelming that they resort to the simplest thing they can think of: the sequential process with perhaps some branching and some swim-lanes. But these come nowhere near giving us an insight into what is really happening in the Theatre of the Third Wave. That's why I made such a fuss, at the start of the book, about instantiation and concurrency being the key to really understanding organizations. That's why they are so central to Riva.

KEY POINTS

When a process architecture 'runs', processes are instantiated.

Process instances operate concurrently.

The dynamism in the world *is* the flux of concurrent process instances. The process architecture expresses the potential concurrent behaviour of that part of the world, in the same way that a process model expresses the potential concurrent behaviour within one process.

We can identify business entities (essential and designed) with their CPs. The world becomes a constantly changing network of process instances and a cascade of responsibilities.

8 Managing the modelling

Provides guidance on running a process workshop and conducting interviews in order to prepare a model of a process, for whatever purpose.

INTRODUCTION

In Chapter 2 we looked at the business of modelling a single process from the 'technical' point of view. But the process modelling activity itself needs to be managed if it is to be successful. This chapter gives guidance on actually doing the process modelling work to ensure that it gets to answers as efficiently as possible. We look at modelling from a procedural point of view (i.e. as a process itself) and, to a lesser extent, from the 'soft' point of view, taking into account some of the sociological, political and people issues. We shall step through a basic process modelling procedure – in particular, a modelling workshop – addressing the soft issues as we go. We'll take a generic view of the modelling process for now, leaving it to subsequent chapters to describe how to adapt it for specific situations: discovery, definition, design, diagnosis and enactment. The four Ds and an E.

There can be no hard and fast rules for the procedure we should follow. It can be affected as much by political and logistical issues as by technical modelling issues. However, here's the basic scheme:

1. Decide on the objectives of the modelling.
2. Brief ourselves by getting an overall picture, no matter how coarse, from a variety of sources.
3. Run one or more interactive workshops of those involved to draw up a RAD that meets the objectives of the modelling.
4. Use other appropriate sources of information.
5. Review, revise and validate the model using other inputs.
6. Use the model.

No surprises there. Remember that we have already prepared our process architecture.

Before we look at each of these steps in detail, there is one topic that we have touched on already but which we need to consolidate: the difference between abstract and concrete process models.

Abstract and concrete models

There is an important choice that we must make at the start of any session: do we need an abstract model or a concrete model, or one that is a mix? When we examined the different process concepts in Chapter 2 – role, action and interaction – we saw how we could choose to model abstractly (concentrating on intent) or concretely (concentrating on mechanism), or both. Let's remind ourselves of the options and then look at two models of the same process: one abstract and the other concrete.

If we are modelling how a process operates *now* we might wish to capture, for instance, the division of labour between people and computer systems. Or we might decide that it is not important initially whether an action involves people or machines, and decide instead to capture the 'essential' content of the process, not worrying how it is manifested physically. Yet again, we might decide that in future we want a particular action to be done by a person, or automated and given to a machine to do, or done by a person with the aid of a machine (such as a database system), and we might choose to model this.

For example, we might prepare a model of how you get a purchase order form from someone who keeps a stock of blanks, fill it in with the details of something you want to buy, pass it to me, get me to authorize the purchase by signing the form, and return it to you. We might represent this process fragment as some actions in our respective roles and interactions between them. Those interactions have a purchase order form as the gram passing between us.

The abstract process might simply be 'You ask me to authorize a purchase,' a single interaction without any gram. What we capture is not the mechanism but the intent. Whether we choose to model a process concretely or abstractly depends on what we are trying to achieve by modelling.

If we are involved in process improvement or re-engineering we might find it useful to:

- draw up the current concrete process model;
- 'abstract' it to yield the current abstract model;
- find a better way of implementing it in a new concrete model.

Remember that we are not thinking here of preparing detailed as-is concrete models, and then deriving complete abstract models, and then complete to-be concrete models. The suggestion is only that by 'moving' in some sense from the concrete to the abstract and back to the concrete we will gain insights. That could mean simply thinking it through in rough sketches or on whiteboards.

This route also makes sense when we are modelling a process as a prelude to providing computer support to some or all of it.

Concrete vs abstract actions

When we draw an action on a concrete RAD we will want to show how the action is done; we shall want to talk about the mechanism. In an abstract RAD, on the other hand, we shall talk about the intent or purpose of an action. So, in Figure 8.1 we see a very concrete action – *Complete form 21b*

FIGURE 8.1 *A concrete action and its abstract counterpart*

– whose name is explicit enough to be a work instruction, but gives us little idea as to what is achieved by taking this action except that a form has been filled in. The figure also shows the action expressed in abstract terms – *Prepare sales analysis for the month* – which tells us what we want to achieve but not how we do it. Of course, we are quite at liberty in our models to give an action a name that tells us both things – *Prepare sales analysis for the month by completing form 21b* – but it is as well to be aware of the two contrasting styles.

Concrete vs abstract interactions

Precisely the same applies naturally to interactions, as illustrated in Figure 8.2 (which we saw earlier in Chapter 2).

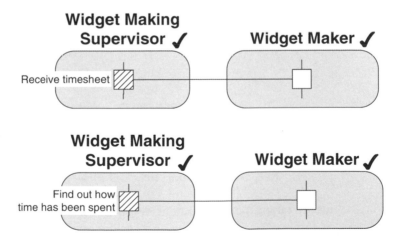

FIGURE 8.2 *A concrete interaction and its abstract counterpart*

When we model interactions concretely we shall see verbs like 'send', 'sign', 'copy', 'pass', 'receive' and 'get' – mechanisms. When we model abstractly we will use words like 'request', 'delegate', 'authorize', 'approve', 'report' and 'agree' – intents.

Concrete vs abstract events

The differentiation between mechanism and intent carries naturally into events. We might label an event as *Form CC received from customer* or as *Customer makes a claim.*

Concrete vs abstract roles

Finally, we can do the same thing for roles, something we discussed a lot in Chapters 1 and 2. When we take a concrete view of a process, we shall be very likely to choose as our roles things recognizable in the organization: posts, departments and computer systems. When we take an abstract view we shall be more likely to identify areas of responsibility without reference to the way they are allocated to organizational entities.

So, in the top part of Figure 8.3 we talk in terms of areas of responsibility called Project Managing and Purchase Approval. In the bottom part, where

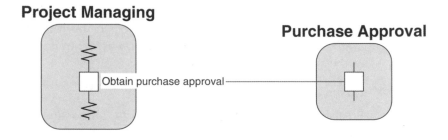

FIGURE 8.3 *Abstract roles and their concrete counterparts*

we are taking a more concrete view, our concrete model shows the job title of Project Manager and the post of Finance Director.

An example

In Figure 8.4 we have a process described in very concrete terms. Real posts and departments appear as roles; actions involve physical objects and physical actions on them; interactions are expressed in terms of the paperwork involved; and there is even a computer system taking part as a role. This would make an excellent RAD for telling people exactly what to

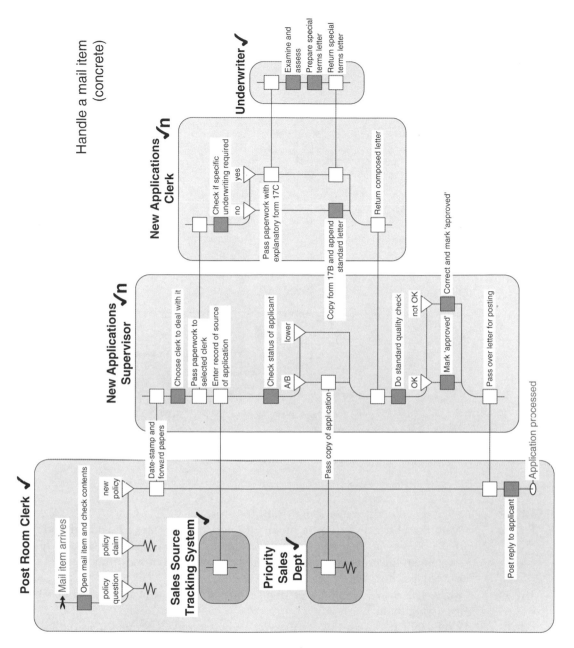

FIGURE 8.4 *A concrete model of the handling of a mail item*

do. But it would be of less use for getting an overall picture of why things are being done.

Suppose now that we rework this model, trying to move towards a more abstract representation of the process. We might draw something similar to Figure 8.5.

Business Process Management

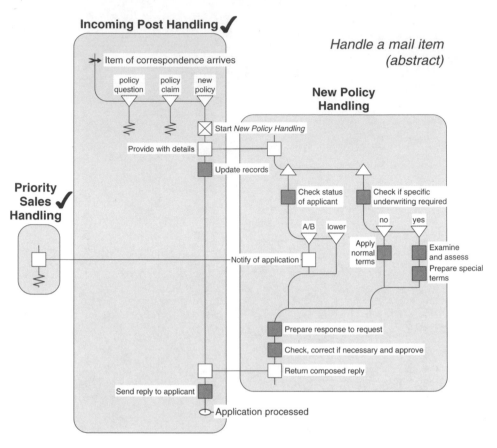

FIGURE 8.5 *An abstract model of the handling of a mail item*

Tutor:	What do you notice about the change?
Pupil:	Well, the abstract version is much simpler. We've removed a lot of stuff that's shown in Figure 8.4 only because of the way they've chosen to implement the process: posts, computer systems, communications mechanisms, paper flow etc.
Tutor:	Right. The abstract model is getting at the *essence* of the process, at what we are trying to do.
Pupil:	Would it be fair to say that it's therefore a better model, because it's simpler?
Tutor:	NO! It's vital to remember that abstraction isn't some sort of *summarizing*. Our aim is to model the process in terms of intent or purpose, rather than mechanism. A side effect might well be that the model is pictorially simpler, because we will ignore 'implementation detail'. But a simpler model is not necessarily a better model. Hiding implementation detail might be the last thing we want to do. It might be precisely our aim to demonstrate that a process is a shambles of crazy activity! We might model all that madness and end up with a crazy RAD – and that will be exactly what we want management to see to prove to them that they need to do something about

it! Showing them a tidy model will not scare them into action. Our second model might be an abstract version, as the first step to simplifying the way the process is carried out. But then we have a different purpose in mind for the model.

KEY POINTS

Before modelling, we must decide if we are working with an abstract or a concrete model.
An abstract process model shows intent.
A concrete process model shows mechanisms.
A model can be a mix of abstract and concrete perspectives.
The type of model must be determined by the purpose of the model.
A simple model is not necessarily a better model.
If it's a muddle, we might need to model the muddle.

STEP 1: DECIDE ON THE OBJECTIVES OF THE MODELLING

The importance of this step can never be overemphasized. Over the years, process modelling has acquired a bad name. This is because all too frequently it has been done badly, has lost the plot, wasted everyone's time and money, and yielded little except a doorstop of pictures. Part of the problem has been the poor methods that have been used: methods that rely on decomposition (which all too easily results in unstoppable modelling), or that provide no clear direction (with random models as a result). The other part of the problem has been that modellers frequently forget why they are doing it, and get stuck in a belief that they are working on *the* model – whereas of course there is no single model, only the model you find useful.

Once we have answered the question 'Why are we doing this?', I suggest we have it printed in red on a large sheet of paper that is displayed whenever and wherever the process is being thought about. We must avoid analysis paralysis. Then, to answer questions such as 'Do we need more detail?', 'Where should we stop?', 'Do we need to cover such-and-such?', just look at that large sheet of paper and rephrase the question: 'Have we enough detail to do what we are trying to do, to answer the questions we have posed, to achieve the goals we have set?'

If we do not know why we are modelling and do not have a clear idea of the outcome we are looking for, or what we want to be able to do with the model, the modelling activity will be slow and undirected at best, and at worst will fail. We are more than likely to end up modelling aimlessly. We have already seen that there is no single model of a process so when we start to model, we need to know what perspectives we should be taking.

Early in this book we identified a number of different reasons for drawing a model of a process. In subsequent chapters, we shall look at these different aims:

- To discover a process or to define it to the degree we want to enforce it – Chapter 9.

- To map the as-is process for diagnosis and to improve it – Chapter 10.

- To design a brand new process from scratch – Chapter 11.

- To provide a basis for the design of an information system – Chapter 12.

- To build an enactable model – Chapter 13.

STEP 2: GET AN OVERALL PICTURE

Now that we know where we are headed, our aim in the second step is to map out the ground; in particular to identify the boundaries of the process model and to identify the perspective or perspectives that will help us get to answers. Our aim is to get ourselves briefed before the group workshops and individual interviews, so that we have a good idea of what we can expect to hear: the roles we can expect to hear about, where difficulties and tensions lie, where the process starts and finishes, and so on.

What is our starting point?

Our first input is of course the process architecture. We might have produced it as part of step 1 when we fixed on the organization that we are interested in, or it might be in place already. Either way, it will have chunked the activity of the organization that we are looking at, and we shall have chosen one or more processes that we want to capture, or design, or diagnose, or improve, or enact.

The architecture gives us first-cut answers to the following questions. Write those answers down and publish them for the process you are dealing with:

- What is the 'organization' that we are concerned with? The process architecture defines the organization in terms of the UOWs it deals with, of course, but let's come up with a simple characterization such as 'All of the Marketing Department's activity excluding TV campaigns', or 'The interface between clinicians and the pharmacy', or 'The work of the group that deals with vehicles abandoned in the street'.

- Are we looking at a CP, a CMP or a CSP?

- How is the process activated or triggered?

- What relationships does it have with other processes? In particular, does it generate new cases of any UOWs itself?

Roughly what happens?

The process architecture only draws a boundary round a process. It tells us nothing about what is inside it, except that we know the sorts of things that will appear in a CMP, for instance. Our next step is to get an appreciation of the contents of our process. My experience is that this can be done efficiently by interviewing someone who has a good grasp of the whole process, even though they might only operate a part of it. They might not be a senior person in terms of rank, but they might well be senior in experience in that organization, having worked in many parts of it. The essential thing is that they should have a broad view. If an organization has previously done some sort of investigation into how things work (or don't work), the person who led that could be a useful starting point. This work can take a couple of days of intensive discussion, using lots of informal pictures.

At some point there might be enough information about the process to take a first cut at a RAD. But remember that this is a private RAD: we are sketching it as part of our briefing prior to getting close to the action.

What roles can we expect?

Early on, we can start to list the roles that we are likely to come across as we get into the modelling sessions. Natural sources are:

- posts from the relevant parts of the organization chart;
- job titles;
- departments, branches etc;
- roles identified in existing written procedures;
- committees, task forces, working groups etc;
- the customer(s) of the process and perhaps their customers;
- suppliers to the process;
- regular meetings.

Using this list we also need to decide whom to invite, to be sure of having the right people in the room to answer the questions we have set. I have a subversive suggestion at this point; unless they have a role to play in the process, do not invite managers ... more of this later.

What terminology is used?

Now is the right time to start a glossary of terms. If, as analysts, we are new to the business we are modelling, we have what is both an advantage and a disadvantage: we don't know the terminology of that business. This is a disadvantage because at first we shall be slow to understand what people are saying when they use terms that have a special meaning for them. This is an advantage too because, in trying to find out exactly what they mean by those terms, we shall start to uncover ambiguities in the organization's views about itself: two groups might use the same word but with different meanings. Such ambiguities can be indicators of misunderstandings or even conflicts.

What do senior people think?

Working our way down through the organization chart will generally be the politically correct approach: involve department heads early on. We should find out whom they regard as authoritative about what goes on in their groups, get their commitment to the use of their staff's time, manage their expectations about possible outcomes etc. What is politically correct and what is necessary to get the facts will determine in what order we speak to whom. This is all bound up with the way in which the larger project – for radical change, incremental change, the introduction of a QMS, or whatever – is itself being handled. Most such initiatives require senior level backing and sponsorship for success, a topic that is adequately covered in the literature on these larger topics and one which I do not address here. But we should be aware of that problem.

STEP 3: RUN ONE OR MORE INTERACTIVE WORKSHOPS

Having been briefed in the process we're interested in and having cleared the way with the appropriate senior people, we now reach perhaps the most important step in the process modelling project: the group session in which we will model the process that we are working on.

But let's first take a small digression here to worry about whether group sessions are a good idea at all. If you are an ethnomethodologist you will believe that the one true way of finding out how an existing process works is to sit and watch it, perhaps even take part in it. We must ask however whether, given the questions we are trying to answer and the challenges we are facing, we are actually interested in pure discovery. If our aim is to design an improved process, we might not be the slightest bit interested in capturing how things are done today: we know the process is wrong – why draw a picture of it? If our aim is to define a process so that we can standardize it – 'this is the way we do things round here' – then we won't want to capture all the ad hoccery of actual practice: we want to produce a work instruction, not a depiction of reality. So whilst we can appreciate the different dimension that ethnomethodology can bring to the depiction of existing operational processes, for the most part that is not our interest.

Our interest *is* in getting a group to engage with their process – for definition, diagnosis, improvement or enactment. If we are not designing a process in a greenfield situation, in other words if there is an existing process out there, we shall aim to get in one room a representative from each of the roles that we have identified during our briefing. It may be the first time those people have been in a room together, so ingrained might be the functional silos of the organization. For the first time they might be seeing how their work fits with the work of others and facing for the first time, from a general perspective, the areas of conflict or stress between groups in the process.

The output of the session will be a RAD. But here is an important point: we shall not introduce the group to the RAD notation. We use it almost without mention, drawing the process as we go, using the symbols we have become familiar with. Our job, as the person at the front holding the marker pen, is to translate what we hear into a picture – a RAD. Each time we use a new symbol – the first part refinement, say – we shall simply make a point just to say 'When we have a number of parallel threads starting I'll draw this ... ' My experience is that any group readily takes to the notation, provided that the analyst actually does the initial drafting for them. It's sufficiently transparent for people to work on their *process* during a modelling session, rather than to work on (struggle with) the notation.

Starting the workshop: laying out the ground

This is not a book on facilitating a workshop. Much of what follows is just a Riva slant on everyday facilitation. However, I like to start by agreeing the timetable: 'Here we are. It's 9.30 am. We will finish by 12.30 am. If you commit to giving me your undivided attention during the next three hours, I commit to finishing this workshop no later that 12.30 am.'

When people enter the room, make sure they find on the wall that large sheet of paper with the purpose of the session on it. Spend some time getting their focus on it.

Next, spend time agreeing the bounds of the process to be modelled. From our briefing work, we already have a good idea of what to expect, of course. But besides scoping people's expectation and thinking, this exercise will help each individual warm to the area and the challenge written up there in red. Questions that help this exploration include:

- **What's the process called?**
  This may sound a stupid question, but it often reveals different perspectives or agendas in the group. To get people to think, I replay their name for it:
  'What's it called?'
  'It's called *Accept delivery of raw materials.*'
  'I visualize some raw materials that have just arrived and you are going to accept them – yes?'
  '... No, we don't always accept them: sometimes we reject them. We're really checking them to see if they can be accepted.'
  'OK. I visualize some raw materials that have just arrived and you are going to check them to see if they can be approved. So perhaps it should be called *Check delivery of raw material*?'
  '... In fact, we check who they've come from: if they come from an approved supplier we accept them immediately; otherwise they are turned away.'

And so on.

Our aim should be to bring the group towards the name that we were led to by the process architecture, **Handle a delivery of raw materials** for instance. If things won't go in that direction then perhaps there is a message there: that the architecture is faulty in some way, or we have the wrong group here, or there are such different viewpoints in the room that we need to back off and deal with the discrepancy.

- **For a CP, what does this process deal with? What does it handle? Or what does it produce?**

  In Riva terms, what's the UOW? We must avoid getting into the input/ output style of thinking, of course: this doesn't help in scoping the process and can too easily have people focusing on the way the process works.

- **How does the process start?**

  In Riva terms, what events trigger action? A CP will generally only yield one trigger but the group might identify triggers for interactions with other processes: **Make a product batch** has to respond to enquiries about progress from management, for instance. A CMP should generate plenty of triggers, as we saw in Chapter 5.

- **How do we know when the process finishes (if it ever does)? What are the goals of the process? What are the possible outcomes of the process?**

  Groups often need prodding to recognize that a process can have more than a successful outcome: rejection, failure, handing off, or escalation can all be alternatives. We must rephrase the question in many ways to winkle them out:

  'How would I know it has finished?'
  'What's the last thing that happens?'
  'Can the process fail in some way?'
  'If that's success, what's failure?'
  'How many different ways can it fail?'

- **Who are the people or groups involved?**

  What roles will we expect to see appearing in the model? People will shout out all the job titles and posts that they can think of. At the edge of the process they will identify larger groups, perhaps entire departments that they have connections with. Write them all up. There may be dozens – I have seen forty. Keep writing. We will probably find that only a handful actually appear on the final RAD, but the list will prove a useful memory jogger.

  Identifying all the roles is not always as easy as it might seem. People think naturally in terms of departments and named individuals and this is a perfectly good starting point. In his work with a leasing company, my colleague Tim Huckvale worked with a group who identified Kate as the person who did such and such – Kate had always done that. Kate's name went on the flipchart.

- **What areas do we want to ignore today? What areas are definitely in the discussion?**

  The process architecture has chunked the organizational activity, so we might know that the processes for certain UOWs are being dealt with in separate workshops and, we only need to go as far with this process as the boundary concerned. If we are dealing with a CP, we know that case management is outside our work. If we are dealing with a CMP, we can ignore how individual cases are dealt with. And so on.

Our prior briefing has told us what to expect to hear and where to probe. We shall put all the answers to the questions onto flipchart sheets and stick them on the walls as reminders whilst we work on the process itself.

For the actual modelling a large whiteboard is essential. Preferably two ... or three. We need a large area to work in and we will be drafting, correcting, changing, and rubbing out a great deal – space and flexibility are key. Do not use flipcharts for the RAD – you cannot rub things off. Above all, do not attempt to use a specialized drawing tool on a PC. It is very distracting, and very inflexible. People must be focused on the process. Icons, buttons, windows and all the flummery of a software application simply divert attention from the real job. If you have a printing whiteboard or a digital camera, so much the better. If the whiteboard is a projected virtual one that you can sketch freehand on, directly into a computer, better still. But please leave your favourite drawing tool at home! If we find we have to sit down at the end of the session and copy the whiteboard contents onto paper or into a laptop, then that is a small price to pay for the complete concentration of the group. Technology distracts.

Doing the modelling

Now comes the moment when the modelling has to start.

If the process has a natural start-to-end flow about it – perhaps it is a CP – we can take advantage of this: we draw the main triggering event for the case at the top left-hand corner of the whiteboard and put it into the role where the process starts. We find out which role that is by asking 'Who notices this?', 'Who first gets to hear that something is needed?', 'Who deals with it in the first place?' We look at our brainstormed list of candidate roles for inspiration. Not surprisingly, this first step can take a worrying amount of time: all the questions about whether this is a concrete or an abstract model, whether we are modelling organizationally or in terms of responsibilities, whether the process starts here or before or after etc, have to be answered. We shall be patient and work through this. These first decisions will set the tone for the entire session, so we must be happy that we have these decisions right before we move on and make more. We should expect to go into the room knowing the appropriate answers, of course – that is our responsibility as leader.

We start by drawing the goal (i.e. the desired outcome) of the process as a state somewhere in some role at the bottom right of the whiteboard. The rest of the workshop is now about filling in the process between these two points. Roles appear as they enter the process, spreading across to the right. Don't start by writing all their names along the top of the board – we shall only use a fraction of the ones that were identified earlier.

If the process does not have a neat flow, things are less easy and it becomes necessary to do the same thing for each of the threads that exist.

Either way, as the RAD develops, we shall draw it and redraw it many times as the group explores the process. Few will have thought about their daily lives in this way, so there is an (enjoyable) element of exploration for them. There will be problems that we have to solve as the modelling proceeds. Just how much detail do we want to get into? Shall we ignore that role's contribution at this stage? Is it sufficient to summarize that set of interactions as just one for now? Do we want to separate those two roles or treat them as one at the moment? Shall we collapse all that activity into just one black box? Should we regard the work going on in that other process as outside our boundary and simply capture it as an external event or two? There are, as ever, no stock answers to these questions. *It all depends why you are modelling: look up at the words in red for an answer.* That said, we must get to the end of the process in the available time: we made that commitment. Going away with at least a rounded – if not 'complete' – model has a value: we can always explore further detail in later sessions, either with the same group or with smaller groups and individuals.

Given that a process will generally have many threads and that they cannot all be explored simultaneously, we shall need to be careful to note where threads still have to be closed off. And it is useful for the group to know that this care is needed so that they are encouraged to point out unfinished threads that will have to be returned to at some time. Simply drawing a 'spring' at the appropriate point on the RAD is usually enough to indicate 'unfinished business'.

Should we look at the 'normal' situation first and then come back and add the exception condition handling and abnormal situations later, or should we try and deal with them all on a single sweep? There is some virtue in the first approach in order to build a framework on which everything can be hung. The danger is of course that exception and abnormal situation handling can easily be forgotten if put off 'until later', and it is often those parts of the process that reveal areas for improvement or suggest the possible use of computer systems to reduce the likelihood of error.

Workarounds – additions to the 'approved' process in order to make it work – are a fruitful source of ideas about what is going wrong and what could be done to remedy things. It is not unusual to hear something like 'Then I go and get approval from the monthly Management Meeting … well – I say that – in fact sometimes I can't wait that long, so I check it out with the Chairman, Brian, and then get it rubber-stamped at the next Management Meeting, otherwise nothing would get done.' That workaround is a clue. Indeed, we shall need to probe for the existence of such workarounds:

- What do you do if you don't get the stuff in time?
- Do you ever get on with that even though you haven't had authorization?
- How long do you wait for that?

As the modelling proceeds, some of the roles brainstormed earlier will find their way onto the RAD. People will start to abstract away from named individuals – 'Well, Mary does do that but she's signing it off in her capacity as Site Safety Officer' – and the roles appear. Quite often roles that are not on the 'main' stream of the process are missed initially. They might only be involved for a single interaction, but of course it can be that interaction that holds things up, simply because it is some form of approval that is required from an 'outside' role: *Get Health and Safety to sign off the risk management plan, Get the plan signed off by QA, Get Finance to agree to the budget.* Equally, we must be prepared to strip out roles that don't materially contribute to the process or its understanding as the modelling proceeds.

This is part of the trick of knowing what to put in and what not to put in – it's a modelling decision. The boundary needs constant validation:

- We've got this role in here – do we care for now?
- Is it worth looking at what happens before this trigger or not?
- Is this really the goal, or is there actually something earlier/later that we are really interested in?

and so on.

Standing pen in hand, at the whiteboard, we have an important task: eliciting the process from the group, getting it onto the whiteboard, allowing the group to own what is drawn and to buy into what is drawn, steering the modelling, and bringing it to a conclusion. The result of their work is that RAD on the whiteboard. But many other important items of information and clues will have cropped up during the discussion and debate, things that are spoken or just hinted at, and for this reason we shall find it very useful to have another person simply keeping a record of things such as:

- avenues that were explored but backtracked from;
- any decisions to ignore certain detail that would need to be picked up later;

- concerns about the way the process works or doesn't work currently;
- suggestions as to how it could be improved;
- situations where errors frequently occur;
- points of stress in the process;
- judgements about the relationship with other processes/departments and their effect on the process under discussion.

People will say a lot of things during the modelling that could act as pointers to inefficiencies, problems, and solutions. We must record these for later analysis. Our note-taker, perhaps more than anyone standing at the whiteboard, has to be sensitive to what is said:

- When these forms arrive, the applicant's policy number is rarely filled in and we end up having to go back to the originator to get the information.
- We generally don't have enough time to handle that fully and we only go back to it when we get a quiet period.
- Couldn't that be sorted out at the weekly meeting rather than waiting for the next management review?

All these signals need to be noted for future use, if we are not going to explore them there and then.

Our aim is that, by the end of the session, the process on the whiteboard is *their* process and the model is *their* model – they have after all drawn it, albeit with the help of the analyst who held the pen. This element of ownership is, as ever, vital for subsequent work.

Closing the workshop

Having made that commitment at the start of the session to finish on time, we shall stick to it. Our aim – whatever happens – is to get to the end of the process, even if we have to gloss over some areas, recognize that we have not adequately dealt with others, and so on. We must cover the *entire* process to some degree. We can always come back and go over it again. But we cannot put it down halfway through and expect to pick the threads up again a week later – not least because we probably won't have the same people in the room.

So, remembering our deadline, we shall also aim to leave enough time at the end of the allotted time to revisit all of the brainstormed triggers, outcomes and roles that we put up on the flipchart. Did we cover them all? If not, why not? Are we clear why not and happy with that? Although we might well have started with twenty or thirty candidate roles, we'll probably find that we only have half a dozen on the RAD – this is quite typical. For instance, the brainstormed list probably had all the individual job titles in the Accounts Department, but – for this model – we only needed to show the role *Accounts*.

The clock says 12.25 pm and we wind up with a final question: 'We are going to take this information away, draw it up tidily and circulate it to you. Do you think we have finished our discovery/diagnosis/design today?' The group needs to decide if its work is done. We should always aim to get them back at least once for a review session. One workshop is not enough. In some cases, four or five have been necessary for a group to get to a final process design that it is happy with, that meets the design goals set for it, or that it thinks it can operate.

As our group disperses, we gather up all the material that has emerged and transfer the pictures and all the other information to a more portable form. We now have to organize what we have heard. Our first task will be to draw up the RADs 'properly' – at last we can use our favourite drawing tool. This inevitably reveals unfinished threads, missing detail, doubts, misgivings, questions about terminology, and so on. We collate these and add all the signals and messages the note-taker heard during the session. We will use these to guide further information gathering, as input to the review workshop to follow and to subsequent workshops that continue the capture, design or diagnosis that we are doing.

KEY POINTS

A good basic procedure for a group modelling workshop is:
1. Arrange to involve representatives from all the likely roles.
2. Ensure the workshop room has plenty of drawing space: whiteboards and flipcharts. And working pens!
3. Get the group together for an agreed period of time.
4. Put the purpose of the overall exercise on the wall.
5. Brainstorm triggers, goals and outcomes onto flipchart sheets.
6. Brainstorm an initial list of roles onto flipchart sheets.
7. Walk through the process from start to end, picking up all the triggers.
8. At the end, revisit the roles, triggers, goals and outcomes and ask if they have all been adequately covered.
9. Record all issues and concerns separately as they are raised.
10. Finish on time whatever it takes.

When is a process model 'complete'?

If we examine an island such as Anglesey or Martha's Vineyard to answer the apparently simple question 'How long is its coastline?' we come up against a problem. If there is a road around the island we could measure its length and decide that that is the length of the coastline. But the road doesn't trace round each inlet (there might be a bridge over it) or round each peninsula (it cuts across the base). To get a 'more accurate' measurement we might decide to walk around the coastline with a pedometer, walking into each inlet and around each peninsula along the coastal path. We will get a greater distance than we obtained from the length of the coastal road. To get a yet 'more accurate' figure we might decide – having much spare time – to run a tape-measure along the edge of the water as we pace the coastal path. We will obtain a yet larger distance. And so on.

This is a good metaphor for processes. There is always more detail if you want to look for it. Whether the detail is useful and justifies the expense of collection, only the process modeller can determine – there is no simple rule that can tell you 'You have finished!' Completeness is in the eye of the modeller. The answer is in those words in red on the wall.

So one session might not be enough. We might need to reconvene the group and revisit those parts of the model that we have not finalized. Our note-taking and post-workshop review will tell us what remains to be done. And it is when we all find ourselves in the room together again that we realize the value of having got to the end of the process at the first session: we have the whole thing laid out in front of us – right or wrong – and we can take a view of it in the cold light of the new day. Having slept on it, when they see it again our group might decide it was all a terrible mistake: 'That won't work – it's too complicated,' 'It still doesn't feel sufficiently responsive to the customer,' or 'Will that really work with the new database that is being put in next quarter?' We will not be afraid of tearing it up and starting again if this should happen. Hopefully, things won't be that bad and the new session will be about filling in gaps, refining where necessary, removing unnecessary detail in places, and general reworking.

By the time we have finished, our RAD might look a tangle, and once again our job is to take it away and rearrange it so that it becomes clearer to read and understand. (Note that I am only going to rearrange it diagrammatically: I am not suggesting that we simplify it in any way.)

STEP 4: USE OTHER SOURCES OF INFORMATION

When we capture, diagnose or design a process with Riva, we work primarily with workshops, an approach which has obvious drawbacks: those involved can tell us untruths, they can forget to tell us about interesting things, they can tell us what they think we want to hear, they might conceal things they don't want us to hear, and so on. How can we deal with this? Where possible, we must deal with it at the workshop itself but there are some other routes that we can use.

Examining existing documents

A document almost invariably gives solid form to an interaction some-where – it is after all a way of collaborating. There are potentially four groups of people involved with a document and they represent roles interacting for some reason:

- *The author(s).* They have some reason for producing it: to inform, to instruct, to report etc.
- *The reviewer(s).* They provide quality control on the document and its contents.
- *The authorizer(s).* They are approving or authorizing the publication of the document for some reason: they are the budget holders; the information is being released in their name; they are responsible for public statements; they are agreeing to certain aspects of it etc.
- *The recipient(s).* They are presumably expected to act on the document. The recipient might receive the document 'for information only' and not act on it, but each 'copied to' role represents a potential interaction, whether or not it serves a useful purpose. When we look at the 'copied to' list we might see a list of job functions or positions – Finance Director, Marketing, QA – or a list of names. In the latter case we have the task of determining which role the recipient is acting when they receive the document concerned.

Documents often record the path that a case has taken through the CP. Good document control keeps a record of the history of a document, making it possible to see what process has been applied to it in reality.

Examining existing documents describing processes

Our organization may well have documented some of its processes in the past. Documented processes can take a number of forms including procedures, manuals, work instructions, and Quality Manuals. Where these exist they will clearly be an important source for us in that, in theory at least, they should describe the process in some fashion. But there is a danger here too in that 'in theory': written procedures and practices are not always followed scrupulously or even at all, especially if they can only be made usable and efficient by 'adapting' them. In a modelling session, people often ask 'Do you want us to tell you how we actually do this, or how we're supposed to do it?'

Highly regulated industries are more likely to have documented procedures. They are also more likely to follow them if there is the constant expectation of an auditor arriving unannounced at the front desk.

Examining existing terms of reference, personal objectives etc

We have seen how a role groups a set of responsibilities within a process. Such responsibilities are often communicated to the incumbent as written terms of reference, objectives, or a job description. These can give us clues about:

- the desired outcome of the work of a role;
- the way the job is to be carried out;

- the things the role is to produce;
- targets that the role must achieve;
- resources that the role can use;
- interactions that the role must have;
- who the role's customers are.

Identifying regular meetings and their purpose

We saw earlier how boards and committees that have regular meetings can play a role, in the sense of having defined responsibilities.

We can represent a meeting simply as an interaction between the roles that are represented at the meeting, or we can regard the group that meets – indeed the actual meeting – as a role in itself. Which view we take depends on whether the group that meets has some responsibility of its own in the process, or is simply a way for the individual roles to get together.

I have modelled a process in which the *Weekly Development Meeting* was shown as a role. The meeting had responsibility for making a joint decision in the process (in theory at least), so we decided to show it as a role. In practice, the meeting was often unable to make decisions, as we discovered when we attended one, because representatives were not always empowered to make decisions on behalf of their departments. As a result the meeting became more of an updating session – an interaction that could be carried out in many simpler, less expensive ways.

When we look at meetings we should ask questions such as:

- Who attends?
- What roles are they playing when they attend?
- Why do they attend? Are they there for reporting, receiving information, authorizing, or taking decisions on behalf of themselves or the group they represent?
- Does the group that meets play a role itself and have its own responsibilities?
- How does the outcome of the meeting get propagated? How does it cause subsequent activity in the process?
- By which roles?
- Does the meeting report its outcome to other roles who did not attend?
- How are they supposed to react?

One way we can answer these questions is to actually attend and observe what happens.

Interviewing individuals

Some items on our 'Issues arising' list might best be addressed by interviewing an individual. If they have not seen a process model before, then we must decide whether walking them through the model is the best way of doing things, as opposed to a simple question and answer session. We need never feel obliged to show the RAD to the interviewee.

Once again I prefer to have two people carry out an interview: one questions while the other records. It can be beneficial to hold the interview at the interviewee's normal place of work. Very often, in order to explain something to us, the interviewee will say 'Let me show you an example . . . ,' reach into their filing cabinet and produce an illuminating document. This has to be balanced against the usual problem of interruptions to the interview and hence everyone's concentration but, overall, interviewing people on their home territory seems most effective.

A two-hour interview is about the most that both sides can take. The interviewee becomes drained, and the interviewers overloaded with information. We should budget about half a day for the two analysts to go over the information gleaned, in particular working it back into the RADs and recording new questions and issues that will need to be referred back to the interviewee, or on to subsequent interviewees or the next group session.

Setting the scene at the interview is key. Time is limited and there is much to cover, so we should spend a few minutes covering a number of points with the interviewee. They go roughly as follows:

- Thank the interviewee for their time.

- Ask how long we have actually got for the interview.
  Although we might have asked for a two-hour interview we are probably starting late, and the interviewee will have subsequently agreed to give someone else the second of our two hours! Agreeing at the outset how long the session will last means that we can pace our questioning and ensure that we cover the key points, rather than wasting all the precious time on smaller issues; and the interviewee makes some sort of commitment to the time that's agreed.

- Outline the purpose of the project.
  The interviewee might well have heard of the project and have some idea of what is going on. We should describe the project overall and then place our activity in that framework. Being open about our motivations helps the interview along. It is generally the case – and we should stress it – that our work is non-judgemental: we are not there to observe and then say 'Aha, there's wastage, that's inefficient, why on earth are you doing it that way?' Our role as facilitators is to bring the organization to these sorts of statements from its own observations and judgement.

- Describe how we are doing our work.
  We want to position this interview in the larger scheme of things. Why are we interviewing people, and this person in particular?

- Say to whom we have already spoken.
  This helps to prevent repetition and to make it clear that either this person is very important and is being seen first, or we have already spoken to this person's boss and hence are here with some authority. There are of course sensitivities here and dangers too: repetition can be a good thing if it reveals differences of opinion about a process, and a different viewpoint often provides new detail or insights. Also, we don't want subordinates to feel that they necessarily have to toe the party line and say what their superiors would want them to say.

- Explain how far we have got.
  How much have we found out so far? What areas do we think we have some grasp of and which do we think we are struggling with? People generally like to tell you what they know, so admitting ignorance at this point encourages them to tell.

- Describe what we are doing now.
  Are we trying to establish the ground? Or do we have a good model already and are now trying to flesh out detail?

- Tell the interviewee how they can help us now in this interview.

- If we plan to use a RAD with the interviewee, tell them, and add that we shall explain the notation as we go along.

This leaves us ready to get to the core of the interview.

STEP 5: REVIEW, REVISE, VALIDATE THE MODEL

Throughout the modelling project our RAD will be under constant change as more information is obtained, other information is discarded, the perspective is altered, and so on. The task is all about using our noses, chasing things to ground, following leads and backing off them. There is no simple procedure for successful process modelling whatever its purpose: discovery, definition, diagnosis, design or enactment. Like any such activity, the skill of analysis is with the analyst as much as with the analytical method.

However, constantly taking the RADs back to the process actors and replaying new versions is clearly a major part of our work. A feedback session is one way of doing this. It naturally involves a group of people who have some stake in the process. It is not unlike the group modelling session, except that the model is now being replayed to the people who originally had a hand in producing it or who, though not originally involved, play a part in the process.

Such a session can serve several purposes:

- It is a way of validating the models that have been constructed.

- It provides a way of letting the process actors work through and explore potential improvements for themselves.

- It can be an important part of change management in a radical or incremental change programme, by providing a communications channel from the change management team back to the organization.

The RAD provides a vehicle for description, discussion, and decision, whatever the situation. There is a risk that a new group will want to redesign the process or change the model for some reason – we must be clear whether they have the power to do that and, if so, on what terms and with what safeguards for the model's consistency.

I always reckon to keep each new version of a RAD as it develops. Small changes can be made to the current version ad lib, as the process is clarified or the perspective is clarified. But when a major change is made – a major realignment of perspective or a major restructuring – we should store the current version (call it version N) and copy it to a new one ($N+1$) on which the major change is made. Keeping old versions serves two purposes. The first is the pragmatic one, that even though a major change feels right today, in the cold light of tomorrow it might not seem such a good idea and it is nice to be able to return to a previous version without pain. The second purpose is that the record of how the RAD changes will itself be a useful teaching tool: we can see how the work went, went wrong, was put back on the right track, diverged, returned, and so on; all this information will help us to understand the modelling process itself and do it better next time.

Quality control of a RAD

As the modelling proceeds, we must maintain tight quality control over the RAD: being tidy helps enormously, especially since we are showing the RAD to people constantly and there is great value in their seeing a consistent notation and usage in what we show them.

As in any language, good style helps communication. I am a stickler for accuracy in representation: RADs provide a concise and unambiguous way of saying things, and there is no point in throwing that advantage away by being fast and loose with the notation – all the rules below are there for a reason. Get into the habit of using them even when sketching a RAD, rather than trying to 'correct' the RAD once it has been captured.

- Label each action in verb-object format.
  For instance: *Prepare the monthly report*; *Classify client request*; *Assemble business case for approval.*

- Identify the nature of each interaction with an appropriate verb.
  For instance: *Agree ...* ; *Approve ...* ; *Delegate ...*

- Label each interaction against the initiating role and word it from the point of view of that role.

  For instance: *Hand over monthly report, Receive classified client request, Approve business case.*

- Label important states so that just the interesting part of the state is briefly described. Describe the state with a sentence in the present tense.

  For instance: *The monthly report has now been prepared, The client request is new and exceeds £10,000, There is no approved business case.*

- Annotate the RAD with a highlighted text block when you need to make a comment about the accuracy of the RAD itself.

  For instance: *What happens now if the application is refused?, What happens to the documentation of a refused proposal?, Regional Manager step omitted for simplicity.*

KEY POINTS

To avoid analysis paralysis, ask frequently 'Have we enough detail to do what we are trying to do, to answer the questions we have posed, to achieve the goals we have set?'

A process model is complete when it is sufficient to address the declared goal.

Complement the workshops by:

- looking at documents that suggest process;
- examining written procedures (whether or not they are actually followed);
- looking at people's terms of reference;
- observing regular meetings;
- doing follow-up interviews with individuals.

Quality control your RADs constantly.

Keep old versions of RADs – you may need them! And they hold valuable lessons.

9 Discovering and defining processes

Covers the practical use of the approach in determining what processes an organisation has, in eliciting those processes (onto RADs), and in the use of RADs in QMSs, tying into ISO 9001.

INTRODUCTION

In the simplest of situations, we want no more than to find out what is going on in the organization. We're in the world of *discovery*: there's a process out there, but no one can quite say what actually happens. People are doing things and perhaps the right results are being obtained. But our individual responsibilities – what we each have to do – are passed on by word of mouth, by rote, or by tradition. The fact that results are produced at all is perhaps a matter of luck rather than judgement.

Once we have been through the process of discovery we shall have a process model that says 'This is how in practice we handle customer complaints.' We're now in a position to give each individual – perhaps for the first time – a helicopter view of how things work and how their contribution fits with that of everyone else – or doesn't: 'This is how your work contributes to customer complaint handling.' At last we're able to share that understanding across a group of people: 'So this is how, together, we handle customer complaints round here.' This will be the first step towards change and small-scale improvement: 'If you could do this, my life would be made much easier; in return I can ... '

As well as discovery, we might want to write up our process as a work instruction: 'This is how we will all handle customer complaints.' This sort of process *definition* can serve several purposes:

- to ensure we continue working together in a way that we all share and understand;
- to ensure that new people joining the organization can get into the swing of things quickly, by following the process that has been laid down;
- to ensure that we deliver the quality our customers are looking for in our product or service by working in a way that guarantees – or at least increases our chances of delivering – that quality;

- to satisfy a regulatory body that governs how we go about our work, by carrying out our processes in the same way, day in and day out, so that we achieve the necessary quality.

In this chapter, we shall look at how best to use the concepts we have developed so far to discover a process that is out there somewhere, and then go on to examine the different ways in which we can present a process definition for the different purposes we might have. But first we need to touch on a subject that will crop up a lot in modelling work: the *boundary* of a model.

The boundary of a process model

When we draw a process model, no matter how much detail we put in it or how 'big' it is, we are always drawing a boundary somewhere, a boundary that says 'For this model, I'm not interested in what happens beyond here.' We are making a modelling decision.

The black-box action as a boundary

The first and obvious boundary in a RAD is the black-box action. By representing an action as a black box we are treating it as an atom, and saying that we are not concerned here with the detail of what goes on in that action. The black box is a sort of lower boundary to the detail about activity that we want to capture. We might wish at some point to examine what happens 'inside' the box – we dealt with this in Chapter 4.

Here are some examples of the sorts of things we might say when describing a single action. Note how they might describe a small, short-lived activity or something very large and very long-lasting.

- Prepare production line for new model.
- Complete handover form signalling acceptance of new production line.
- Agree pricing with dealerships.
- Issue Press Release.

The interaction as a boundary

Similarly, an interaction is an atomic thing: we have no detail about what goes on when the interaction takes place. It just happens at the appointed states in the participating roles and that's an end to it as far as the model is concerned: another lower boundary of detail. Again, we have chosen not to look more closely in this RAD.

As with an action, the 'size' and duration of an interaction can vary enormously:

- Supplier demonstrates equipment to Potential Customers.
- Supplier and Customer negotiate price.

- Supplier and Customer sign agreement.
- Supplier maintains Customer's equipment for ten years.

The trigger as a boundary

Any trigger is clearly a boundary too. Suppose we have the event *Decision made to carry out an audit* as the trigger of a thread in a role. This says that the thread starts when this event occurs, and that – on this model – we are not interested in where or why or how it occurs, only that it occurs.

Pre-existing role instances as a boundary

We can consider pre-existing role instances as a form of boundary to our model. If we draw the role *Task Force* and mark it as having a pre-existing instance, we are saying 'However it came about, there is a single instance of the role *Task Force* when this process starts.' An instance might be there because it's a fixed post in an organization, such as *CEO*, or because some other role in some other process – we care not what on this RAD – has created the instance in order, say, to activate this process.

Boundary definition by omission

Finally, we implicitly define the boundary of a model simply by omitting things. The most common example of this is where we only draw those parts of a real-world role that are relevant to the process we are looking at. Figure 9.1 shows how we might wish to indicate that, for certain items of

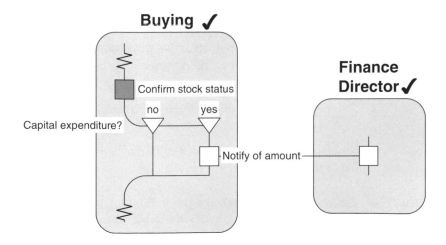

FIGURE 9.1 *Omission as boundary*

expenditure the Finance Director needs to be told, and to that end there is an interaction with the *Finance Director* role. But we have chosen *not* to show just what the Finance Director does with the information – this is outside our interest for this RAD. Clearly, the Finance Director does more in life than is suggested by this RAD, but everything besides accepting notification of capital expenditure is outside the boundary chosen for this model.

DISCOVERING A PROCESS

Let's begin by looking at how we go about discovering (uncovering?) a process. As ever, we shall start with a process architecture. The important thing about a Riva process architecture is that we don't need to know anything about how things are being done. In other words, we don't need to know how things are done before we structure our understanding of how things are done. The process architecture is deduced only and wholly from an analysis of the business the organization is in, its 'subject matter', which we examined in Chapter 6. By going through the EBE and UOW analysis we reach an architecture which says 'If you are in this business, which has these units of work, then you must have these processes with these dynamic relationships.' That's a very powerful place to start the challenge of discovering how individual processes in the architecture are done. The chunking we achieve ensures we build on sound foundations.

So, when we start on the discovery of a particular process within the process architecture, we already have a name for it and we know where it starts and where it ends. Let's remind ourselves of the basic scheme from Chapter 8:

1. Decide on the objectives of the modelling: here we are concerned with discovering the process.

2. Brief ourselves by getting an overall picture, no matter how coarse, from a variety of sources.

3. Run one or more interactive workshops of those involved to draw up a RAD that meets the objectives of the modelling.

4. Use other appropriate sources of information.

5. Review, revise and validate the model using other inputs.

6. Use the model.

What adjustments do we need to make to this general approach? We shall listen in . . .

Tutor:	Firstly, no managers are allowed to this workshop!
Pupil:	They won't stand for that! Why don't you want them there? They feel exposed already because the process isn't defined, which presumably is their fault. And even if they have been delinquent, they'll want to know what is happening out there.
Tutor:	The problem is that the people doing the work might be – shall we say – shy about revealing how they do it. So by all means, let's interview the managers as part of step 2, finding out what the managers think is happening or expect to be happening, or even know is happening. But we are more likely to get the 'truth' from the coalface, where the dirt is!

Pupil:	So at the workshop, we'll have in front of us a group who should not feel intimidated and who will tell us how things really happen, good or bad, appropriate or inappropriate.
Tutor:	Right. Secondly, we'll draw a concrete model.
Pupil:	If we want to understand the process shouldn't we be thinking abstractly rather than worrying about footling detail?
Tutor:	We want to know what happens in reality. We might choose to stand back a little to understand the purpose of the actions and interactions. But it won't help to get too abstract, too early.
Pupil:	OK, but if we want to extract what is really going on, what sort of questions should we be asking?
Tutor:	Easy: 'Do you really?', 'Why do you do that?', 'Why in that order?', 'Does everyone in that role do it the same way?', 'Has it always been done like that?', 'Where do things go wrong and what do you do then?', 'What's behind doing it like that?', 'Why does the same person do those two rather different sorts of things?'
Pupil:	So we're standing in front of some 'coalface' workers; we know we are going to draw a process model. But where do we start?
Tutor:	At the beginning of course! Suppose we're working on a CP. When we looked at the general scheme for running a process modelling workshop, we drew the main triggering event for the case at the top left-hand corner of the white-board and put it into the role where the process starts. And we drew the goal or commonest outcome of the process as a state, somewhere in some role at the bottom right of the whiteboard. Once the trigger is on the board we can just ask 'And then what happens?' and we draw what we hear. There will be discussion; there will be questions about whether we are getting too detailed or not detailed enough. As the analyst, we must help the group make that decision at each step. Once a new blob is on the model, we ask the same question: 'And then what happens?'
Pupil:	How do we handle the fact that processes are very rarely just a simple sequence? There will be all sorts of threads.
Tutor:	Yes, there will. The trick is not to forget that we are leaving a trail of unfinished threads behind us – just use that little spring symbol to say 'We have to come back here and pick up this thread.' When we started the workshop we brainstormed all possible triggers and all possible outcomes, and we shall need to check through those lists for more processes before we finish.
Pupil:	If this is true process discovery, there will be lots of surprises and it will be hard to do that pre-workshop briefing you talked about.
Tutor:	It will. It means that we have to listen very hard at the workshop itself to what people are saying and not to let anything go. It's why I suggested that there should be a separate note-taker present to catch all the points that might slip away as discussion continues.
	We'll still also use all the other means we can to find out what is happening. All the methods we listed in Chapter 8 are at our disposal: interviews, observation, documents, terms of reference, personal objectives … anything that might suggest how things are actually being done. It all gets assessed for the model.

Choosing perspectives for communication

If we are modelling to communicate, the scope and perspective of our model need to be chosen so that the model tells people what they need or want to know and omits everything else, or perhaps relegates it to 'the edges' of the model. One strength of a RAD is that we can decide at each point of the boundary how we want to draw it: with a black-box action, with an external event, or with an interaction, as we saw above.

At each point in the modelling we can ask 'Is it important to us to know this or to know beyond this point?' or 'Does it help our understanding to get into this area?' In some instances we might say 'definitely not' and simply terminate that thread; in other instances we might say 'We don't need to know the detail but it would be useful to have some context' in which case we might 'summarize' the process beyond the area we are interested in. Again, on a RAD it is quite simple and natural to mix the level of coverage to suit your purpose.

We might have more than a 'neutral' aim of communication. We might want to scare. The process is a mess and we want people – management in particular perhaps – to realize that it is a mess. The keynote of our perspective here is *honesty*. Whatever it looks like we will model it: if it's a muddle, we will model that muddle. Drawing a tidy model of a messy process makes no sense if our aim is to reveal the mess.

KEY POINTS

When modelling a process for discovery:
- keep the managers away;
- keep the model concrete;
- keep challenging;
- draw whatever is interesting and helpful;
- draw muddle if there is muddle.

PRESENTING A PROCESS DEFINITION

Suppose we want to use our model as a *formal* description. Perhaps we plan to use it as some sort of work instruction, in a Quality Manual for instance; perhaps it is to be used as the keystone of a Process Standard or a Standard Operating Procedure. We want to *prescribe* how a process is to be carried out: 'This is how we do things around here.' As with process discovery, we shall start with a process architecture to chunk the organizational activity and decide what processes there are. The question then is 'What perspective do we want to take of the process given that we want the model as a definition, to guide/instruct/require of people?'

Choosing perspectives for prescription

Let's look at the features we shall expect of our model:

- Our model will need to be sufficiently precise and detailed for us to be able to 'dictate' what we expect of people where we want to dictate to them, and to leave room for discretion where we want discretion to be used. Warning: most people prescribe too much and their process definitions stray into areas where people could and should be left to their own devices, discretion and initiative.

- Our RAD might need to be precise and testable enough to allow independent auditing of the way the process is actually being carried out in order to check for conformance. Remember that 'precise' does not mean 'excessively detailed'; 'testable' means there is a way of deciding afterwards whether it really happened that way.

- We shall need to take care that we do not say more than we need to. If we are going to be audited against the defined process, then the definition should only say what we want to be audited against. It is all too easy to over-egg the definition and then find we have prescribed more than is necessary for our purposes, and the auditors are demanding we do things that we don't always want to be held to.

- The roles that appear on such a RAD will probably be *concrete* roles – either posts or groups – and the descriptions of actions and interactions will involve concrete mechanisms. Conformance is easier to establish if definitions are close to the real world.

- Independent verification or validation plays an important part in highly regulated industries, and in a RAD we have the mechanism – through roles – for showing this explicitly. We will expect to see separate roles that have responsibility for authorizing actions, approving actions, and signing things off, and we will expect to see interactions with them that express the execution of that responsibility.

Using RADs in an ISO 9001 context

The international standard ISO 9001:2000 (*Quality Management Systems – Requirements*) is commonly used as a specification by organizations intending to manage the quality of their products or services through the management of their processes. In support of the QMS approach, the standard says that it:

> ... encourages organizations to analyse customer requirements, define the processes that contribute to the achievement of a product which is acceptable to the customer, and keep these processes under control. A quality management system can provide the framework for continual improvement to increase the probability of enhancing customer satisfaction

and the satisfaction of other interested parties. It provides confidence to the organization and its customers that it is able to provide products that consistently fulfil requirements.

So, in an ISO 9001-compliant QMS we can expect to find the relevant processes defined in some way. RADs offer a useful way of presenting such definitions. *The TickIT Guide*, which is used for ISO 9001 certification in the software development arcna, recommends RADs for modelling processes in a QMS (see *The TickIT Guide*, details at www.tickit.org).

Figure 9.2 shows ISO 9001's process view of quality management in an organization. Customers, in whatever form they appear, have require-

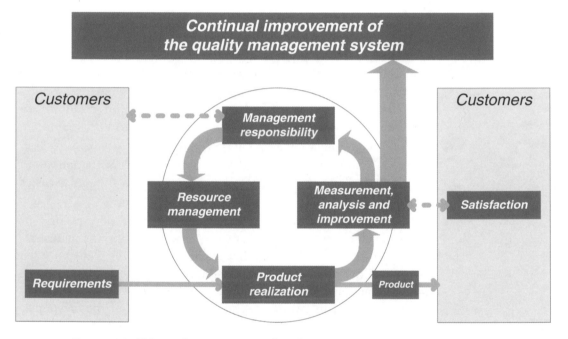

FIGURE 9.2 *ISO 9001's process view of quality management in an organization*

ments which are realized by the organization's processes and presented to the customer in the form of a product or service, hopefully to their satisfaction. 'Product realization' will be where our CPs will typically sit. 'Resource management' will be where our CMPs will typically sit. In Chapter 10 we shall touch on the question of where measurement sits so that process analysis and improvement can be done.

In an ISO 9001 context we will expect to see an emphasis on the control of the process, especially quality control and corrective action; in other words answers to questions such as: 'At what points in the process are checks on quality carried out?' and 'If a fault is discovered in the product or service what action is taken (i) to correct it, and (ii) to ensure it does not happen again?'

I have seen RADs presented in a number of ways as part of the definition of a process:

- A RAD on its own.

  One of the great secrets of RADs is that captions to symbols are drawn *next* to small symbols and not inside big symbols. This is such a small thing and yet it is a massive modelling convenience. Whoever decided that the lozenge should represent decisions in flowcharts deserves a hundred lashes: it's the most impractical shape you could choose. In a RAD the structure of the process is carried by the ordering of the blobs and becomes a great deal more visible. The captions are free to take up as much space as needed, alongside them. We are at liberty to put a small paragraph against, say, an action blob if we find it useful. Thanks to this remarkable property, a well-constructed RAD has little need of additional supporting material and yet it is very compact.

- A RAD with roles expanded in text.

  There are, however, those who simply cannot work with a picture – something to do with left- and right-handed brains perhaps. An organization that was preparing a QMS to be used by many hundreds of people prepared the procedures in their Quality Manual in the form of a RAD followed by text structured by role. Each action, interaction and decision in a role was elaborated in text in a section following the RAD. Someone carrying out a role could therefore see everything about that role in one place, both in the text and in the RAD.

- Text supported by a RAD.

  I have also seen procedures in the form of traditional text description 'supported by' a RAD. This is generally unhelpful. It is notoriously difficult to express a complex concurrent process using something as serial as natural language. I once took a 24-page text description of a process and reworked it as an A3 RAD, with practically no loss of information but with a huge gain in visibility ... and during that reworking a host of problems were revealed that the text hid.

- A RAD on an intranet with hot links to ancillary material.

  This is much more interesting. A RAD fits naturally in the world of web pages. Where supporting text is required, hot links to separate pages carrying that text can be provided from the relevant blobs on the RAD. Or links can lead to pro formas to be used, data systems to be accessed, and more.

- An enactable RAD.

  This is perhaps the ultimate in process definition: the process model is loaded on a process enactment system that is able to 'drive' that process and all its actors. This is a major topic and an important one in the world of BPM and we must leave it to Chapter 13.

KEY POINTS

When modelling a process for definition/prescription:

● make the model a concrete model, especially if it is a work instruction;

● say things just once;

● only say things you want to be held to;

● use the RAD as the hub of the definition and hang other material off it.

SUMMARY

Figure 9.3 shows the general scheme that we use for discovering and defining existing processes. As with any process work, we start by

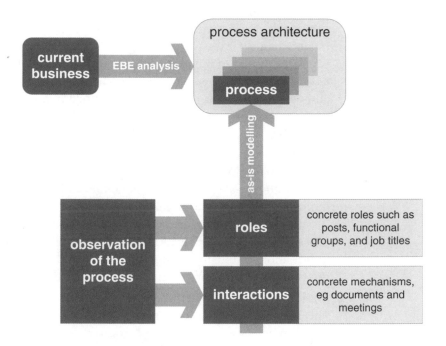

FIGURE 9.3 *The general scheme for discovering and defining processes*

preparing a process architecture to get that ideal chunking into processes. We can then take the process we are interested in and start work on that, confident that we are starting from somewhere appropriate. 'Observation of the process' covers all the different ways that we might decide how a process is being done.

10 Analysing for process improvement

Covers using the approach at both the architectural and the process level for asking questions about processes and their performance, and for driving tactical process improvement.

THE IMPORTANCE OF SPRING-CLEANING

Two three-letter acronyms have dominated the world of business processes over recent decades: TQM and BPR.

This is not a book on TQM or BPR, both huge topics that have the concept of *process* at their heart and that also contain many soft (but difficult) issues such as the management of change, visioning, motivation, culture and ethics. Our concern here is to look at how Riva can provide answers to process questions for both. In particular, I want to highlight how the concepts that Riva uses give us ways of looking for possible improvements.

Over time, our organization and its processes become convoluted. A process that started out simple and clean has, somehow, become complex and messy. The business itself has changed or the business environment has changed around it, and it has not changed its process in response. Let's look at some of the ways that such complexities can arise.

'We won't let that happen again!'

Bad experiences all too readily lead to extra twists being added to a process. A change might be made to plug a loophole, particularly if a mistake has at some time caused trouble, such as financial loss. Suppose that one day, we ship some equipment to a customer but the spares arrive late. Perhaps that has happened before but without any major repercussions – the spares have caught up a little bit later. This time the equipment failed soon after arrival and the spares were needed quickly ... but weren't there. Our customer sustained a significant loss of business and we had to compensate them. 'We're not going to let that happen again' vows our senior management, and extra steps and checks are added to the process to plug that possible loophole. Of course, on average each shipment is now held up that bit longer, just to ensure that the earlier mistake is never repeated.

Every time there is a process failure a new check, double-check, extra approval or sign-off is added to the process. Gradually the barnacles cover the once sleek hull of the ship and friction builds.

'We won't give them another chance to mess us up!'

Functional groups can become self-protective. Suppose group *A* provides materials to group *B*. And suppose *B* starts to get trouble with poor quality arriving from *A* and, as a result and too often, has to return material for reworking, having wasted its own materials and resources working on poor inputs. Group *B* is very likely to put in extra checking procedures at the interface with group *A* and add elaborate handover and fault-reporting procedures, in order to protect themselves in the future.

Now everything is checked, everything is signed for, faults are recorded and reported formally, and there are procedures for tracking faults and cross-charging for the costs of faulty work. More complications for what should be a simple handover process. None of that new activity adds value, only cost.

'The world has moved on!'

The nature of our business can change. A process that was adequate under one set of market conditions can become quite ineffective when those conditions change and, if the change occurs slowly, we shall add workarounds bit by bit to try to overcome the increasing inadequacy of the old process.

Suppose we are in the business of making fitted kitchen furniture. Some years back when life was slower and customers less demanding, our order-filling process relied on a simple flow of manufacture, with orders and changes to orders being dealt with on a cyclic basis: on the first Monday of each month the production team would look at the recent orders and adjust the production schedules. The whole thing satisfactorily revolved around that monthly cycle.

But our end-customers started demanding quicker delivery, and some stores started promising faster delivery to certain customers. Some orders couldn't wait for the monthly production meeting and so had to be handled as exceptions: 'short circuits' and 'fast tracks' were added to the simple monthly cycle to handle the exceptions. But as time has passed and things have become more hectic, we have added more and more such workarounds to the main cycle and the workarounds have become institutionalized, with the result that our monthly meeting is now more about finding out what is happening and handling the big bush fires rather than planning. Planning is effectively being done out there in the workarounds.

We need to replace the old cycle-driven process, which is now unwieldy and unresponsive, with an event-driven process, one which is clean and responsive. Instead of piling things up to be handled, we need to deal with things as they come in and make the necessary changes to production schedules on the spot. The current role structure and the interactions in it, as defined by the monthly cycle, could continue to exist in order to *monitor* the process from a managerial standpoint, but the day-to-day *control* of the process needs to become the focus.

'We're in another world!'

Have we changed what we do? As well as teaching courses to full-time students, do we now offer remote-learning packages to part-time students? Have we decided not to outsource chemical assays to a service company any more, and instead to bring that function back in house? Have we passed all our invoicing to a finance house? Have we added generic pharmaceutical drugs to our product line as well as under-patent drugs?

In each of these situations we might have tinkered with existing processes to deal with the change in business. But knowing how we can characterize business through its EBEs and then go on to derive a process architecture in terms of its UOWs, we immediately see that we really should be thinking of changes at an architectural level rather than at a process level. We may find we need entirely new CPs, CMPs and CSPs. Or we may need to delete some processes. Relationships between our processes may have changed and we might have made those changes simply by bodging.

If our business has changed in its UOW characterization, we must start by reworking the process architecture.

STARTING POINTS

There are several questions we need to answer before we can start work on the process architecture or any individual processes.

What improvements are being sought?

Our process or organizational improvement work will not reach its goal if we do not know what the goal is. It might be about reducing cycle times so that we can get to market faster, or close the sale more quickly, or answer a customer's query more quickly, or be able to offer a faster response than our competitors. It might be about reducing the cost of dealing with a repair, or putting a new PC on an employee's desk, or recruiting a new member of staff, or preparing a bid for some work, or managing the flow of call-centre calls.

How we decide on our improvement goal is a topic outside the compass of this book – any book on process improvement from a high-level point of view will provide guidance. (Chapter 11's guidelines for designing a process are also relevant when improving a process.) So let's assume that we start with a clear idea of what improvement we are looking for.

Where shall we focus our changes?

Big changes or little changes?

At one end of the scale, we can completely rework the fundamentals of how we do our business: in Riva terms we would, for instance, be examining designed UOWs and asking whether they are strictly necessary, and whether a different approach using different UOWs would streamline things. We collect money for goods by issuing invoices. But why? *Invoice* is a designed UOW. What happens if we delete it? We decompose our software development projects into work packages based on the stages of a development lifecycle – what would happen if we were to decompose them into work packages based on a functional breakdown of the system? (Software developers will recognize this as the step that was taken in moving from the SSADM development approach to the DSDM approach.)

In the middle of the scale, we might reallocate responsibilities amongst roles to reduce the number of handoffs in preparing an invoice, thereby reducing the work involved and the time it takes to issue one.

At the other end of the scale, we might reduce the number of times a rework action in **Handle an invoice** is carried out by improving the layout of the invoice we send to customers.

We can look at process improvements right across this scale.

Intent or mechanism?

We can concentrate on *intent* or *mechanism*.

Are we doing the right things (intent)? The process might be highly efficient in what it does ... but it does the wrong thing.

Are we doing the things right (mechanism)? It might do the right thing ... but it does it inefficiently.

We shall need to decide whether we want to work with an abstract model (which is about intent) or a concrete model (which is about mechanism).

To-be or as-is?

We can start straight into modelling the process as it should be after improvement, the to-be model. If we know the as-is process is bad, why draw a picture of it? Let's just start designing a new process. In some cases, we can safely ignore the existing process and our process modelling workshop will design the new one from scratch.

Alternatively, we can start by examining the process as it is and look for improvements. If we expect the changes to be at a more detailed level, an as-is model might be the right place to begin. In yet other cases, we might not even know what is going on currently, and have to find out through an as-is model before we can start to look for opportunities to improve the process. And we should not assume that the current way of doing things is all bad – there might be very good reasons for certain things being the way they are and it is good to understand them and not make a mistake by reworking them.

Where should we be measuring?

Measurement is a central part of BPM. We need measurement to know how we are performing and to get indications of where things are not as they should be. Our process architecture gives us a vital separation between the process for dealing with one thing, and the management of all the things currently in progress. In particular, the time that elapses between the activation of a CP and the moment it reaches the desired outcome is easier to identify: it is obvious where we should insert measurement probes. Moreover, the CMP is where such measurements can be collected, and trends detected and responded to.

KEY POINTS

Before we begin we must decide:

- what improvements are being sought;
- whether we can consider big changes and/or little changes;
- whether we are concerned with intent or mechanism;
- whether we need to start with the to-be or the as-is process;
- where measurement comes into play.

ANALYSING THE PROCESS ARCHITECTURE

Let's begin at the beginning. Do we have the right process architecture? Do we have the right designed UOWs? In a sense, this is the central question of BPR: are our processes aligned to our business and our customers? So our starting point is to build the process architecture for the part of the organization we're concerned with and to examine that.

Focusing the searchlight

Our first step is to decide what we mean by 'our organization'. This team? The sales force and its customers? Production Engineering and R and D? The branch network? We saw in Chapter 6 that we can adjust the searchlight according to our purpose. Since the EBEs characterize the organization we are talking about, when we discuss which EBEs are UOWs for 'our organization' we shall be moving the searchlight until it illuminates the right thing. When we have finished we shall have a process architecture that covers precisely the organization we want to talk about. As a rule of thumb we can expect to extend across the boundary by one process simply to recognize the points where our processes interact with those outside our organization, or its customers and suppliers.

The process architecture should – if we have done it properly – only include processes for *essential* UOWs. We now add *designed* UOWs: UOWs that we have decided to have and use in order to do our business – in reality, of course, we may have spotted some of these when looking for the essential ones and have bracketed them out. Now we can add them back in but – importantly – we distinguish them from the essential ones. We can think of this as adding mechanisms to an essential model. We saw in Chapter 6 how we can choose to include smaller and smaller UOWs, as if we are seeing fainter and fainter stars between the brighter stars in the night sky. We must ask 'Is this UOW too small to be worth bothering with?' Our answer will be driven by whether we think there is mileage in questioning its reason for existence.

Deleting designed UOWs

Knowing which UOWs are essential and which are designed, we now examine each designed UOW and ask 'Are we happy with this designed UOW as the way of achieving our goal?', or 'What happens if we do away with this UOW? What could we do to achieve the same goal but without incurring the work?'

This sort of questioning makes us realize that *Invoice* is not the only way of getting payment from a customer: we could simply require them to pay electronically into our account against the delivery note when the goods arrive. Instead of batching delivery requests and treating a batch as a UOW, doing all the deliveries in the batch together, what would happen if we simply dealt with each delivery when it arrived?

Outsourcing and insourcing UOWs

We can also ask 'Would this UOW be more cheaply/quickly provided by someone else as a service to us?' – the classic outsourcing question. If we have let the boundary of our process architecture cross into our supplier's domain, we can ask the opposite question: 'This UOW that we currently get from an outside service, could we do it more cheaply/quickly if we did it ourselves?' – the classic insourcing question.

UOWs are precisely the things that we can consider out/insourcing. Roughly speaking, if we decide to outsource a UOW, its corresponding CP, CMP and CSP all go to the outsourcing company. If we insource a UOW, we must take those three processes back.

Task force or service function?

In Chapter 4, we saw the difference between a UOW that is provided by a service and one that is provided by a task force. Suppose we run a product company. For each product we prepare a user manual which needs to be printed. Clearly, *User Manual Print Run* is a UOW which will have a CP and a CMP. We now have a choice.

We can set up a group that will do manual printing as a service. This group will operate the processes **Do a User Manual Print Run** and **Manage the flow of User Manual Print Runs**. We shall demand that any product team that needs a print run for its user manual must use that service. Our goal is to make cost savings by having one set of facilities dedicated to printing manuals. But it does mean that a product team must knock on the door of **Manage the flow of User Manual Print Runs** and ask for a print run, and then take its turn in the queue.

An alternative is to allow each product team to arrange its own printing. It must have its own **Do a User Manual Print Run** and **Manage the flow of User Manual Print Runs**. The case management is probably nil. The team must set up a task force to do the printing. It does not need to join a queue. But it does have to solve the problem of facilities itself, perhaps by going to a local print shop or perhaps by setting up and staffing its own facilities. Our goal is to stop the printing of manuals becoming a bottleneck in getting products onto the market, and we are prepared to pay some extra for that. In essence, we are building a process architecture in which those that use the outputs of a process operate that process – a classic re-engineering strategy.

Asking this question – shared service or not? – becomes very easy with a Riva process architecture.

> **KEY POINTS**
>
> Using EBEs, we draw a boundary round our 'organization' to include just the areas where we think change is necessary.
>
> Starting with those EBEs we prepare a process architecture.
>
> We examine *designed* UOWs and ask if we are happy to continue with them in the future.
>
> Task force and service functions are interchangeable.
>
> We judge a UOW 'too small' for the process architecture if we are content to leave how it is handled out of the study.

PREPARING A PROCESS MODEL FOR ANALYSIS

There is no single way to prepare an as-is process model as a basis for analysing an individual process. Instead we have several options.

Our first model should be a concrete model. 'This is what we do.' 'These are the mechanisms we use.' 'This is what happens on the ground.' As we prepare the model we shall see the problems, some staring us in the face. Above all we need an honest model. We want to see it as it is, warts and all. In fact it's the warts we are looking for. We do not want a cleaned-up version.

We might then choose to derive an abstract version from the concrete model: 'This is what we are trying to do.' Comparison with the concrete model will give us further messages. The convolutions of the process on the ground will be all the more apparent when we compare it with the simple thing we are trying to achieve.

If we plan to take a quantitative view of the process, we shall need details about the duration of the individual actions and interactions and about the amounts of resource they use: people and people's time, quantities of input materials and any other resources needed. In some cases we might already have this information from measurement, or we can estimate it. If we find a significant spread in the values we measure, we might find it useful to record that spread in some way: perhaps a sketch graph of the probability or something along the lines of 'It normally takes two to three days, never less than one and in rare circumstances up to six.' Once again we simply annotate the action or interaction on the RAD with whatever information we wish to keep.

We want to know where people don't use the prescribed process and why they don't; we want to know where the workarounds have become necessary and perhaps even institutionalized; we want to know where rework has become a way of life; we want to know where the big hold-ups occur and why ... all the dirt. This is one of the reasons it can be important to exclude managers: we want a safe environment in which the unpleasant facts can emerge.

ANALYSING THE PROCESS MODEL

If a process is in trouble or inefficient, the symptoms will often be recognized and understood by its actors. No process model can of itself provide the answer to a process problem: it can only act as a sort of searchlight on the process. The purpose of the process model must be to *reveal*: to reveal the process, the roots of its problems, and potential ways of attacking the trouble. Sometimes a RAD can reveal the nature of the problem and suggest a solution very quickly; something in the RAD is like a flashing light saying 'Here's your problem.' In other situations, revelation comes more slowly, perhaps as the right – revealing – perspective is homed in on.

We shall want to analyse the process from two points of view: *qualitative* and *quantitative*. To set the scene for a description of the sorts of quantitative and qualitative analysis that are possible on a RAD, let's first look at the different styles of process improvement that are possible.

There are four ways in which we can improve matters in a process:

- by 'point-wise' improvements to individual actions or interactions;
- by 'flow-wise' improvements;
- by restructuring roles;
- by realigning the organizational structure and the process structure.

In any given BPR or TQM programme, some mixture of these will be used. The first three are generally the domain of the TQM disciplines, concentrating on incremental change and incremental improvement; the last two are where BPR looks for radical change, in addition to the architectural level we looked at earlier in this chapter.

We take these in turn before considering the sorts of analysis of a RAD that would lead us to answers at these four different levels. It is also worth reminding ourselves once again that which of these levels we want to consider will very much determine the perspectives – and hence the RADs – we elicit and draw.

Making point-wise improvements

This is the finest level of granularity in process improvement. Here we are concerned with increasing the efficiency (use of resources) or effectiveness (reliability and quality of result) of individual actions in the process.

We might choose to help the individual carry out an individual action by giving them tools to do their work. In an office environment, helping the individual to do their work increasingly means using information technology with personal productivity tools such as spreadsheets, organizers, and word processors for example. We might improve the way that interactions are carried out by providing email, better-equipped meeting rooms, videoconferencing facilities, online discussion groups, or even an arbitration service. Which actions and interactions we choose to concentrate on when searching for point-wise improvements depends on where we are seeking the benefits of improvement.

Reducing overall cost

Any action consumes resources and some interactions do too. Suppose our concern is to optimize our process in its use of resources and hence its cost, and suppose we have annotated the RAD with the resource usage of each action and interaction. There are some straightforward analyses that can be carried out on a RAD:

- Work on the big hits first. Look at each action and interaction in turn to see which have significant resource usage and which therefore might yield the biggest savings.

- Rework equals waste. Look for case refinements concerned with checking for poor quality. Look at the frequency with which the thread that deals with faulty material or errors in previous work is followed. Trace back through the process and identify where that fault is introduced. What can be done in that action to reduce the likelihood of poor quality? And can the fault be detected earlier in the process so that the cost of rework is reduced? These are all traditional TQM-style questions which the RAD can help us answer.

- Duplication equals waste. Do different roles do the same thing?

- Look for opportunities for error. Remove them.
- Examine the potential impact of supplier inputs. Poor quality will mean increased cost to fix. See what can be done to improve them.

If you are familiar with activity networks for planning purposes, you will know that one of the important ideas when working with them is the *critical path*, that sequence of activities whose total duration is the duration of the process. If an action on the critical path takes ten days longer then the whole process takes ten days longer. The critical path determines the overall duration of the process. We saw above how a RAD can be annotated with the duration of each action and interaction. It can be a simple manual activity to find the critical path. The presence of loops in our process makes the situation more complicated of course, but the principle remains. In complex cases, assuming the loops can be compressed or ignored and all but the longest thread on each case refinement discarded, the (reduced) RAD can be transferred to a traditional project planning tool which will find the critical path automatically.

Actions – especially those on the critical path – can be externally focused, i.e. focused on the customer of the process; or they can be internally focused, i.e. supporting some internal function that is not directly relevant to delivery to the customer. We can go through the RAD classifying each process element (action or interaction) under one of three headings:

- It delivers value directly to the client of the process (mark these in green).
- It delivers value only to the organization: internal reports, measurement, delegation, approval etc (mark these in blue).
- It represents the existence of waste in some form: quality control, handling exceptions, correcting mistakes, apologizing etc (mark these in red).

Ideally, only green process elements should be on the critical path. Blue elements should be moved off the critical path if necessary. Red elements need to be eliminated, of course, though this will require change to the other parts of the process to make it right-first-time.

Shortening cycle time

All actions and interactions take time. We might be seeking a reduced time-to-process for an individual case: we want to get the result – the service or the product – to the customer as quickly as possible (this might be different from optimizing the throughput of an individual worker). Critical Path Analysis (CPA) will be a key tool here:

- Look at each action and interaction on the critical path. If we can reduce the duration of any of those we shall, by definition, reduce the duration of the whole process (up to the point, of course, where another path through the process becomes the critical path).

- Look for the actions and interactions on the critical path whose duration has the greatest variation. In some situations, the perceived quality of a service can be increased as much by making the service *reliably* of a certain duration, as by making it shorter. 'I don't mind if it takes three days providing I can be sure it doesn't take any more.'

This works quite well except for situations where the process has loops or repeated action in it, or threads that are only traversed under some circumstances, for instance for rework. Standard project planning tools cannot deal with loops and case refinements, so we must remove them first if we intend using such tools. We can do that in one of two ways:

- By adding some 'overhead' to the part of the process that, on occasions, needs to be redone.

  Suppose we have a reviewing cycle that is repeated until the thing being reviewed – a document say – is deemed acceptable. We might, for the purposes of CPA, simply assume that the reviewing cycle is always done twice and replace the loop by two reviewing threads in sequence. Or we might collapse the cycle into a single action.

- By treating alternative paths in some proportional way.

  Suppose we are processing forms and suppose type A forms take six days while type B forms take twelve days. About two-thirds of the forms are type A. We might say that the processing of type A and B forms takes eight days on average, thereby removing the case refinement.

This is not very satisfactory. We would prefer to leave the full process structure in place and deal with it intact with loops and alternative threads. The quantitative analysis approach of *System Dynamics* can help here (see for instance Roberts *et al.*, 1983). In System Dynamics, a process is seen as a set of 'flows' of material between 'stocks'. The flows form a network which can include feedback loops and alternative paths for materials. PC-based tools for animating such models allow the flow rates to be specified as formulae and, in particular, to be made dependent on each other, on stock levels, on external variables, on the passage of time, and on the time of year or time of day; those relationships can involve statistical probabilities, so we end up with a model that can be animated, allowing us to determine the cycle time as a probability distribution itself: a much richer representation, especially if we are trying to understand the variability of the time a process takes. Such a model also allows the process's long-term behaviour to be explored: 'Is the process stable over time?', 'What happens to the throughput and cycle time over the seasonal rush?'

A RAD can be converted into a System Dynamics model by turning states into stocks, and actions and interactions into flows. Case refinements become split flows whose rates sum to one, whilst part refinements become joint flows in which the material is replicated on each flow. In practice, the RAD must be greatly simplified first as it contains a great deal of detail that does not need to be carried into the quantitative model.

Discrete simulation models permit similar quantitative representation of a process in terms of flows of 'stuff' from one action to another, but that stuff now takes the form of discrete objects with attached properties which can be handled differently by an action according to their attributes: green widgets are packaged in tens, red widgets in twenties.

To complete a quantitative model we will generally need to collect information about other 'influencing factors' that affect the quantitative behaviour of the process such as:

- the rate at which cases arrive for processing;
- the time of year (the rate at which work arrives might be seasonal);
- staff morale, which in turn might affect ...
- staff productivity;
- the numbers of staff available to carry out different actions;
- the availability of tools, machinery and other resources needed;
- consumer confidence;
- interest rates;
- weather patterns;
- population movements.

We can start to see that a quantitative model deals with factors which are strictly outside the sort of model we have prepared with Riva. Moreover, experience has shown that a quantitative model is most often beneficial if it is kept at a fairly high level. So we should not expect to find quantitative solutions from a qualitative model, nor qualitative solutions from a quantitative model. Whilst there is some overlap between, say, a RAD and a System Dynamics model, we are better off thinking of them as complementary tools, each with its own things to tell us. That said, a process architecture can be a better starting point for a Systems Dynamics model as it operates at a more appropriate level than a RAD.

Making flow-wise improvements

So much for dealing with individual actions, interactions and decisions. Given a set of roles and responsibilities, how can we improve the *flow* through the process? What changes can we make to the *order* of actions and interactions within a role in order to reduce the overall case processing time, or reduce resource requirements?

Increasing parallelism/concurrency

One obvious approach to reducing the overall elapsed time that it takes to process a case is to increase the overlap of activity, especially where this reduces the length of the critical path. This is an approach well known to those who plan projects using activity networks.

To reduce the overall elapsed time of a project, the planner looks for the critical path through the project and looks for ways to increase the amount of concurrent activity, so that actions that were once done sequentially are now done in parallel. The RAD equivalent is to move from the process fragment on the left-hand side of Figure 10.1 to that on the right-hand side.

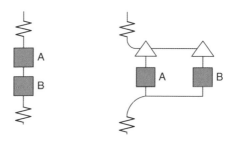

FIGURE 10.1 *Increasing parallelism in a role's actions*

The assumption is of course that *B* does not depend on *A* and that there are actors available to do *A* and *B* concurrently.

The effect of this is to change the elapsed time of this fragment from the sum of the times it takes to do *A* and *B* to the maximum of those two times.

Our inspection of the critical path might suggest that there is no need for *B* to be on it at all and moving it off will be our immediate response.

It is a good discipline to ask 'Do we really have to do these things in sequence?' whenever we see more than two black boxes in a row.

Applying the 80:20 rule – from generalists to specialists

Does every case that goes through a process need to undergo the same processing? Does every purchase order need to be seen by the Finance Director? Could we limit the ones needing the FD's approval to those over a certain value? If we can do this, we can reduce the *average* time it takes for a case to be processed, if not the maximum. It might look as though we are complicating the process, and indeed we are, but the effect can be positive, assuming that the extra decision-making requires very little time. We might replace the top process fragment in Figure 10.2 by the bottom fragment.

We can take this further. Rather than routing both simple and complex cases through the same case workers, thereby requiring all case workers to have the same high level of skill so that they are able to handle any case of any degree of difficulty, we can consider filtering out difficult cases at

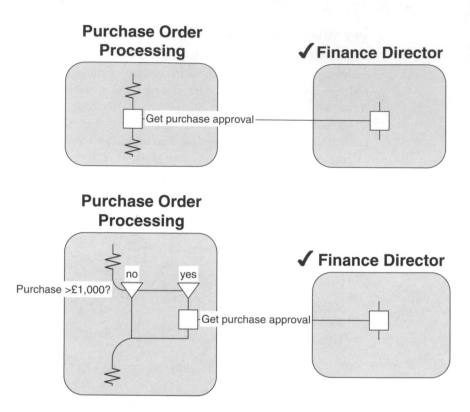

FIGURE 10.2 *Complicating a process for improved average speed*

some point during their processing and passing them to a smaller number of expert personnel. The personnel who handle the run-of-the-mill cases no longer have to have the same degree of skill and could therefore be a less expensive resource. Once again the process becomes slightly more complex to incorporate the filtering, but the benefits could outweigh the costs. The equivalent in RAD terms is shown in Figure 10.3.

As ever, we can also ask the opposite question: rather than having specialists, and filtering mechanisms and handoffs to serve them, can we not have generalists who offer a one-stop-shop?

Planning for success

'Planning for success' is a technique that we can use where shortening elapsed time is of paramount importance and where we can countenance wasting resource if the potential time gains are great enough. A product development process is a typical example. Getting to the marketplace earlier can mean increased product life and earlier revenue flow. A business might consider the possibility of doing work that has to be thrown away if development is abandoned, if it means it can get the product to the marketplace earlier in the cases where development is successful. A

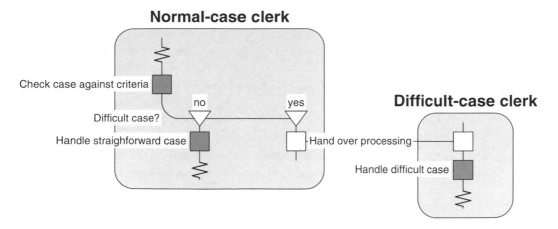

FIGURE 10.3 *Splitting cases by difficulty*

pharmaceutical drug company might well build a production plant for a new drug before they have obtained regulatory approval to sell it.

Suppose that the 'sensible' way of doing things is:

1. Do action *A*.

2. Decide whether it is worth continuing.

3. If it is, do action *B* and continue with the case; otherwise abandon the case.

If we decide at step 2 that it is not worth continuing with the product, we have not wasted effort doing action *B* to no avail. The time it takes to get through this is, of course, the sum of the time to do action *A*, the time to make the decision (typically small in comparison) and the time to do action *B*.

If we plan for success, we start actions *A* and *B* at the same time. When *A* finishes we make the decision. If the decision is 'go' we let *B* continue, otherwise we chop *B*. This is equivalent to replacing the left-hand fragment in Figure 10.4 by the right-hand fragment. (Remember that an action can

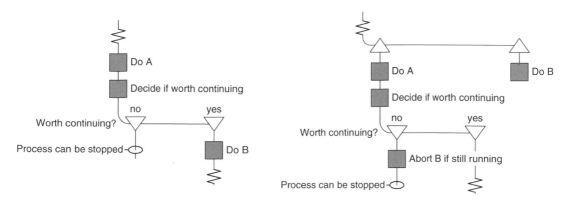

FIGURE 10.4 *Planning for success*

be terminated for a number of reasons: in this case either *B* finishes naturally because it was allowed to run to completion, or it was aborted. How we proceed at the end of the part refinement depends on whether *B* was aborted or not.)

The upside of this change is that whenever the decision is 'go', we save the elapsed time of the shorter of doing action *A* and doing action *B* – we get to market that much faster. The downside is, of course, that, whenever the decision is 'no go', we have wasted whatever has by that time been spent doing *B*. Whether the upside is considered bigger than the downside is clearly going to change from one situation to the next, but the RAD provides a way of exploring the possibility.

The pharmaceutical industry offers many examples of the potential benefits of planning for success. Each extra day that a successful drug is on the market can mean a considerable amount on the bottom line; this can justify risking wasted effort on actions that prove to be unnecessary.

Checking for coherent flow of 'stuff'

We can annotate the actions and interactions in a process model with the 'stuff' – materials or information – they need and the 'stuff' they produce (dare I say their inputs and outputs). Given that annotation, we can check that the flow through the process is *coherent*, i.e. that stuff gets around the process in one way or another, from the roles that produce it to the roles that use it. It is not unusual for the workarounds that we find in a process to be there simply to cope with the inadequacies of the formal process, when it comes to moving stuff between the roles. 'Why is that interaction there? Well, I normally end up going back to the originator because, for some reason, the information asked for in section 5 of the report is rarely enough for me to do my bit.' Those workarounds, as is often the case, give us clues for improvement of the process.

I've lost count of the number of times that someone in a process modelling workshop has said 'I always wondered why you sent me that stuff – I've always binned it.' With any luck they will then say something like 'If it came with the summary sheet, it would save me having to reconstitute the figures myself every time, which is all I want – can you do that easily?'

We can ask such questions as:

- Is anything generated but never used?
- Are all received grams used by the receiving role?
- Is all stuff needed generated?
- Is all stuff generated before it is needed?

Catching faults earlier

If someone makes a mistake somewhere in the process it might not be found until later on, and then correcting can involve unwinding things and tracing back to the source to get the fault corrected – complications in the process and delays in the processing.

Earlier fault detection can reduce the likelihood of faults getting through to later stages and can reduce the cost of correcting them. We can check the RAD for places where faults are detected and see if the detection can be moved to an earlier point, nearer the source.

Look for iteration, where a passage of process is repeated until something is right, or of the desired quality. See if the number of iterations can be reduced by inserting more up-front preparatory work, for instance.

> ## KEY POINTS
>
> **Examine the critical path.**
> **Look for places where there is rework, workarounds, or hold-ups.**
> **Consider ways to improve the effectiveness and efficiency of individual actions and interactions.**
> **Consider ways to shorten timescales by adjusting the flow of activity within roles.**
> **Look for ways of catching faults earlier.**

Restructuring roles and interactions

As we map the process we reveal a structure that is the result of perhaps decades of change in people's job descriptions, in the way the organization is structured, in the business of the organization, in how the organization likes to treat its staff, in the principles the organization holds to, in the use of technology, in unionization ... a host of factors for change. The precise content of each role – particularly where it is defined as a post or group – will not be wholly rational if we view it dispassionately. But from our role-centred RADs we can look for ways of rationalizing the structure of roles, in particular by moving actions between roles, combining roles, and so on. We might explicitly try to reduce the number of interactions that are necessary to make a process work, and this would typically mean restructuring the roles and what they do.

In summary, the RAD becomes a way of challenging both the culture and the allocation of roles and responsibilities (both fixed and dynamic) to different roles in the organization.

How can we detect these sorts of possibilities? With abstract models.

The process as pizza

If we could design our process without worrying about who was to do what, or how the process would work with our particular organizational structure, we could imagine coming up with the perfect process in the form of a nice circular pizza. It would be a tidy, simple structure with only the absolute essentials. But in real life we have to take that nice circular pizza and divide it up between the different people we employ in the different parts of the organization. When we cut up the pizza and pull the sections apart we find we have a mess of strands of mozzarella on our hands. The more pieces we cut, the more the strands.

Each strand is an interaction. It does not add value. It is only there because we have cut the pizza that way. And the obvious observation is that by cutting the pizza another way – by changing our organization so that the pizza pieces are fewer when it is divided – we can simplify things and have less of a mess.

To understand that key relationship between the process and the organization – hierarchies, management structures, job titles etc – we need to use concrete and abstract process models.

Suppose we run our as-is modelling session using a concrete model. We are in effect showing how the many responsibilities in the process are divided up amongst the posts and departments. Division induces interactions. Now suppose we examine the concrete model and draw the equivalent abstract model, where we only concern ourselves with intent and we ignore mechanisms. The difference between the two models will point up where the real-world partitioning into posts and departments has unnecessarily complicated the simple – and 'natural' – division of responsibility.

Moving responsibilities between roles to reduce handoffs

Take as an example the process **Purchase an item of capital equipment**. We have found out that the CEO is asked by the Purchasing Manager to approve purchase orders. To label the role *CEO* as shown in the top part of Figure 10.5 would give a correct model of the concrete process, of the ways things are actually done.

But of course, the CEO box on the RAD does not capture everything done by the CEO, only how the CEO contributes to the process under consideration. We might then look at that contribution and try to characterize it in some way. We might decide that what the CEO is actually doing is giving approval on behalf of the Board, taking into account the cash position of the company and the priorities for contending calls on that cash; the CEO has access to the knowledge of what else is going on in the company and the priorities necessary to make the decision. It just so happens that the CEO is the post currently designated to make that decision. But in our desire to speed up decision-making we might be prepared to move that responsibility around.

By concentrating on the responsibility – the abstract role *Approving Large Cash Outflow* in the bottom part of Figure 10.5 – rather than the current holder of that responsibility, we allow ourselves to think more

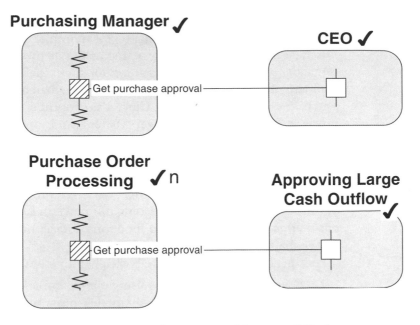

FIGURE 10.5 *Can we move this responsibility?*

radically about whether, for instance, *any* Director could undertake the decision if the necessary background information were available to them too. This gives us a candidate for restructuring. We might end up with something like Figure 10.6 in which we have also allowed the Purchasing

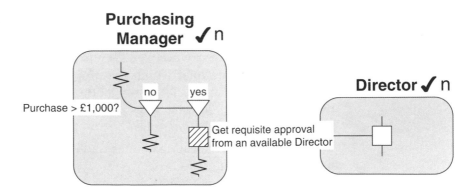

FIGURE 10.6 *Fewer and easier handoffs*

Manager to commit to purchase orders up to £1,000 without getting approval.

Let's take a look at what happens in the Reception area of the building. It might be tempting to regard the work of the staff in Reception as a process in itself: there are clearly some people there doing things during the day, the same people in the same area of the building. But this would be to fall into the trap of associating a group of people with a process. In fact, of course, the Reception staff undertake a variety of roles that contribute to a variety of processes in the company. For example, they contribute to our **Purchase something** process by acting as Goods Inwards for certain sorts of goods, receiving deliveries, signing for them, determining who in the company the delivery was for, notifying them, matching delivery with the purchase order, and so on. They also contribute to our marketing processes insofar as they greet visitors and operate the telephone switchboard. Moreover, they contribute to the training processes in the company by gathering the names of those wishing to attend internal seminars, by arranging food, AV equipment, rooms etc. We might label these roles as *Small goods receiving*, *Visitor greeting*, *Telephone answering* and *Seminar logistics handling*. Doing this makes it easier for us to see how our mapping of such roles onto functions – Reception – can be analysed and perhaps changed.

Let's generalize this. There are basically four steps:

1. Draw up the as-is process using concrete roles.

    We make the roles match the actual posts or functional units that exist in the organization. We are going to assume that the organization will remain the same – the same posts and functional units – but that we want to explore how we can divide the process over them in a way that reduces the number of mozzarella interactions – handoffs – that are necessary. Remember that every handoff is a potential waste of resource, a potential delay, and a potential buffer and conflict point.

2. Deduce the abstract roles underlying the process.

    The actions and decisions have, over time and for a variety of reasons, ended up being the responsibility of the various posts and functional units shown. Once we have the as-is concrete model, we look for the abstract roles within the process. We can then either redraw the process in its fully abstract form, or simply mark them on the as-is RAD by drawing lines around actions and decisions, grouping them into abstract roles.

3. Identify ways of re-allocating actions and decisions in the abstract roles to the concrete roles.

    We decide where actions and decisions could be moved between roles in order to reduce the number of interactions necessary. This is not a mechanical process: it requires experimentation and it can also require the organization to rethink some of its policies, particularly in areas to do with delegation.

4. Define the new concrete roles.

    Depending on the criteria we are using to choose the 'best' re-allocation of actions and decisions, we leave ourselves with a new, restructured process in which we might still have the same roles as we started out with, but now with their responsibilities changed.

Alternatively, we might have created new posts that combine responsibilities more efficiently.

The key to all this is that the concrete and abstract models are helping us to look at the relationship between the process and the organization.

Relaxing/strengthening approval and authorization

A RAD is excellent for revealing the approval and authorization mechanisms that the organization has put in place. They can all be questioned, with a view to either strengthening the mechanism or relaxing it. 'Does the Finance Director *have* to see all purchase requisitions?', 'Would it be better to introduce the requirement for senior management approval at this point, rather than waiting till further down the process?', 'Should this sort of situation be escalated to a higher management level than it is now?'

We can recognize five levels of relationship or delegation between manager and managed (see the excellent book by Oncken, 1987):

- Wait until you are told.
- Ask what to do.
- Recommend what should be done.
- Act but advise at once.
- Act on your own, reporting routinely.

We can ask what the current interactions tell us about the current relationships, and ask whether they are over- or under-strict.

Specialists to generalists – the 'case worker'

When we divide a single task over two people, we generate a need for interaction between them across which the task will flow. When a case or gram moves from one role to another via that interaction, we will often find a *buffer*. If the respective roles process their own cases – UOWs – according to their own cycles, some way is needed of 'decoupling' the cycles at the point where they intersect. That's a buffer – it's what happens in a CMP: requests arrive and go into the planning melting pot, possibly to be batched with others, possibly to be put at the back of the queue. Concretely, it can be folders accumulating in an in-tray, unread messages in someone's electronic mailbox, or all the other ways we have of piling up work to be done. Buffers introduce delays, break the flow of processing, and make tracking and monitoring difficult.

The flow of work for a single case can appear very simple, perhaps of the kind shown in the left-hand process fragment in Figure 10.7. Apparently, each role makes its contribution to the processing of a case and passes it on to the next role down the production line – rather like a bucket chain at a fire: each person in the chain turns to their left, grabs the bucket and swings it to their right. Provided everyone is synchronized it works fine. The bucket (i.e. the case) moves smoothly down the chain. Most CPs are

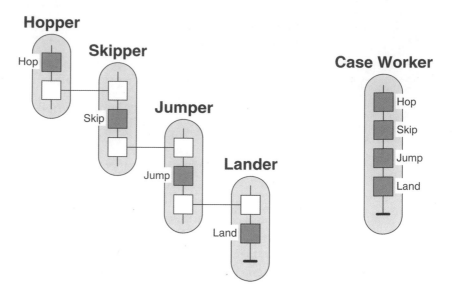

FIGURE 10.7 *From specialist stream to case worker*

not like this; the roles take different or varying amounts of time with their contributions and different numbers of people are put in to deal with each stage to even out the flow. In other words, we end up introducing case management at each interaction and each CMP introduces its own buffer to smooth the flow: as a result buckets pile up between certain individuals and some people then hand on three buckets at a time.

As each role (often a department) tries to optimize its performance, so the overall flow can start to become uneven as the compartmentalization works against optimization of the overall process. Many traditional production-line industries have, for many reasons including motivation, moved away from the bucket chain approach and introduced 'case workers' who take responsibility for the entire handling of a case, as suggested by the right-hand side of Figure 10.7. In automobile manufacture for instance, a production 'cell' might take a car through from chassis to final inspection, working with it right down the line.

Is the problem with the case or the case management?

A CP itself might run more or less well, but if the corresponding CMP is ineffective – or even totally absent – then individual cases can conflict with each other, priorities can be poorly understood and ineffective, and people might have to make do to get by. One would expect to see ways of resourcing individual cases, managing changes in the case load, handling sudden changes in the priority of individual cases and changes in the nature or content of individual cases. If these are not adequately dealt with, there could be trouble.

That said, the word 'adequately' is an important one: it can also be the situation that a CMP is *too* cumbersome and is unnecessarily bureaucratic. The modelling will help us to decide whether things are under- or over-managed.

Is everyone doing something useful?

In the most extreme cases, we might observe roles which have few or no actions of their own and which seem to be only third parties in other people's interactions. These roles may be redundant, adding no value and only slowing things up. I have seen a role which seemed to sit between two organizational units and passed stuff between them. It was quite hard to see what value was added en route and some hard questions were asked about that role. Figure 10.8 suggests the sort of thing we might see.

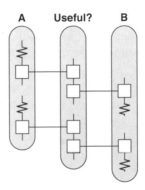

FIGURE 10.8 *Is this intermediary role adding anything?*

Analysing interactions – handoffs

The interactions in the process are there because our roles – organizational units – need to collaborate, and the way we define our organization determines the way the collaboration works: we get more or fewer mozzarella strings. By looking at the interactions we can therefore get some clues as to how the organization might or might not be getting in the way of the process.

(Auramäki *et al.*, 1992) examines office communications using discourse analysis, and its analysis of 'illocutionary acts' is a useful way of examining interactions in RADs.

We can think of an office as a communication network which creates, maintains, and fulfils commitments. In a RAD, commitments happen in interactions, so we should find it useful to analyse these to see what sorts of commitments are being made. (One interaction might involve several separate commitments.) If we refer to the role that initiates the interaction as the 'speaker' and the other roles as the 'hearers', we can identify five types of commitment:

- *assertive* – the speaker is noting an actual state of affairs;
- *directive* – the speaker is asking the hearer to do something;
- *commissive* – the speaker is committing to do something;
- *declarative* – the speaker is bringing about a new state of affairs;
- *expressive* – the speaker is expressing attitudes or feelings about the state of affairs.

An interaction carrying an *assertive* commitment would be one in which roles are being informed about the completion of some action: 'The plan has been issued,' 'The invoice has been paid,' 'The budget has been finalized.' We can ask 'Do the hearer roles really need to know? Is *B* interested in the assertion?'

When the commitment is *directive* we would expect to see the hearer role doing something as a result (or else why are they being told?), and at some later point reporting back in an assertive commitment to the speaker role, to confirm completion. We can check this: is the interaction *Request report on fault statistics* matched by a subsequent interaction *Deliver report on fault statistics* in the other direction? We can also check that each hearer is able to carry out the requested action (i.e. has the authority and resources, knows what is to be done, and is willing to accept instruction from the speaker). And finally, we can check that the speaker has the authority to give these instructions.

When a speaker makes a *commissive* commitment (i.e. promises to do something), there should be a later interaction in which the speaker makes an assertive commitment to the hearer, confirming completion of the promise. We can check that this interaction occurs: is the interaction *Agree to supply necessary resources* matched by a subsequent interaction *Confirm necessary resources are available*? And we can check that the speaker has the authority to make the commitment in the first place.

An interaction carrying a *declarative* commitment might be hard to distinguish from one carrying assertive commitment, but we would expect in the declarative case that all the roles involved would see this interaction as a trigger to get on with some new action: they already know *what* to do; this interaction is giving them permission to proceed. We might find the interaction *Production can now start*, or *The specification can now be relied upon*, or *Approval to proceed has been given*. We can check whether all roles know what to do.

(*Expressive* commitments are outside the sort of modelling we undertake with RADs, so we shall not examine them further.)

In Chapter 2, we looked at how any interaction could be regarded as a 'conversation for action'. Any interacting pair of roles in our process model will be engaged in such a conversation, possibly several. We can use the template in Figure 2.38 to analyse the relationship between two roles by identifying the separate interactions and mapping their components onto that framework. Are there conversation components that are missing? If so, should they be present? If they aren't, would the process be improved if they were added?

Finally, we can ask the following general questions to test whether we have the right interactions between roles and, possibly, whether we have the right roles:

- Is there a pair of roles with a mass of fine interaction? This might indicate a poor division of activity between the roles or a confusion over objectives.

- Are there roles which have the same type of interaction with many other roles? This might reveal a pervasive function that should be dealt with separately.

- Is the concrete form taken by an interaction 'long' in some sense? It might be inefficient.

- Does an interaction have a buffer in its concrete form? Buffers slow up interactions. The existence of a buffer might reveal a designed UOW and its CMP that we had not recognized before.

- Does all the checking, authorization, referring back, copying for comment, input, or approval etc help in the achievement of goals?

KEY POINTS

Draw the as-is process using concrete roles.

Deduce the abstract roles underlying the process.

Identify ways of re-allocating actions and decisions in the abstract roles to the concrete roles.

Examine the interaction structure to see what it tells us about the division of responsibilities across concrete roles.

Consider restructuring roles to improve the interaction structure of the process.

Check for buffers and the hidden UOWs whose flow they manage.

CASE STUDY 5

In organization Q, a sizable mechanism had evolved over the years to deal with requests made to a service group in the organization. Dealing with these requests – which involved making and delivering specialist goods to a hard and fast timescale and specification – was a top priority for the organization and hence for the service group. When the requesters said 'Get these supplies to this location on this date,' it had to happen. To make life that little bit more interesting the requesters were, for perfectly valid reasons, in the habit of changing the details of their orders at any time and making new requests at short notice. Arms and legs were broken by the service group to make sure the requests were correctly satisfied, and invariably they were. But the stress was increasing and it was not clear that the mechanism, or the people, could stand a strategic change in the business that would lead to larger numbers of smaller requests. What was to be done to reduce the stress? Point-wise improvement – increasing the productivity of the actions required to make the supplies – was not the issue. Some other solution had to be found.

We modelled both the CP (how an individual request or change to a request was satisfied) and the CMP (how the stream of requests was handled, prioritized and so forth).

The main issue was the fact that satisfying a request required contributions from four main teams, each struggling to solve difficult technical problems to fulfil its contribution. Each new request or change to a request meant getting all four teams to rejig their schedules whilst keeping all previous requesters happy.

In the old days, when life was slower, this 'negotiation' of changed schedules could be dealt with by one of the regular meetings held by different committees: there were four altogether, dealing with the management of the process from the day-to-day tactical level to the long-term strategic level. But as time pressures had increased people had found that the formal process was too slow, and during the modelling of the CMP we found many workarounds where people would try to sort out a solution and then get it signed off at the next appropriate meeting. This became evident from the RADs which showed, buried in a mass of workarounds, the original four time-driven cycles. What the RADs also revealed was a mass of bilateral interactions by which the four teams attempted to negotiate changes to their separate schedules, outside the formal meetings.

These insights into what was happening led the service group to look for a new way of running the case management. What was needed first of all was an event-driven process rather than a cycle-driven process: things could no longer be held up until the next cycle – they had to be dealt with as and when they came in. The need for a quick resolution was now the norm not the exception. The UOW was not the *Monthly meeting*: it was the *Request* and the *Change to a request*. Indeed, the latter had not been recognized at all and was treated as an irritation. Moreover, the negotiation of schedules between the four teams needed to be made the responsibility of a single new body (role) that would replace all those bilateral interactions with a single negotiation, recognizing that many bilateral negotiations led invariably to whirling around in circles ... and stress.

The change of process pattern and the introduction of a new role simplified the case management considerably, resulting in less stress, continued satisfaction of the requesters, and increased ability to handle future workloads.

SUMMARY

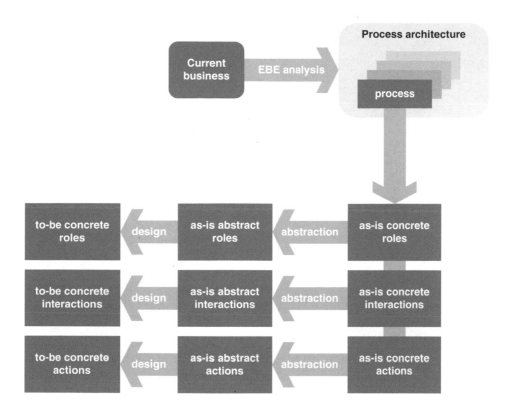

FIGURE 10.9 *The general scheme for process improvement*

11 Designing a process

Covers the design of a new process architecture and new processes starting from a blank sheet of paper.

INTRODUCTION

So far, we have been dealing with the situation where there is a process out there and we want to unearth it, understand it, communicate it, standardize it, analyse it, or improve it. Or several of those. In this chapter we look at how we can use Riva in a greenfield situation where there are no processes at all, or there are some processes and we want to add one or more new ones. In some extreme cases we may have a process in place but it is so awful we have decided not even to look at it and to replace it from scratch.

For simplicity, we shall assume from here on that we have a truly greenfield site. The other situations can be dealt with by appropriate adjustments to the greenfield approach.

As with process improvement in Chapter 10, this is not a book about running a process development *programme*, more a book about how Riva's concepts can be exploited in such a programme. So we shall touch on the programme aspects but only to connect Riva into them. (For that larger context, you might look at Burlton, 2001).

Here's an overall plan of our process design project as it might appear in the larger process programme:

1. Prepare a process architecture.
2. For each process we want to design:
    (a) Set the ground rules for that process.
    (b) Run a process design workshop.
    (c) Validate the design.
3. Revisit the process architecture to ensure that the fit we decided on at the outset has worked.

PREPARING THE PROCESS ARCHITECTURE

This should no longer be a surprise: whatever the situation, we shall start with a process architecture. If we have a new organization – be it a company or department or whatever – we shall go through the procedure described in Chapter 6: characterize the organization through its EBEs, filter those down to UOWs, establish the dynamic 'generates' relationships between the UOWs, translate the resulting UOW diagram into the first-cut process architecture, and reduce that to give the second-cut process architecture. Initially, there will be no designed UOWs and hence all the processes – case, case management and case strategy – will be to do with the things that are the essence of the new organization. As we start to design those processes, we might decide to create new, designed UOWs that will be part of the way we decide to do the business of the new organization. These will invoke new CPs, CMPs and CSPs which we can also design.

GETTING THE PROCESS DESIGN GROUND RULES IN PLACE

With the process architecture in hand, we can confidently work on each of the processes. Of course, we can't expect to knock them down one by one, sequentially. But in what follows we shall assume that we are addressing one process in isolation.

We must be certain we are headed in the right direction with the right terms of reference. We shall ask the following questions:

- Are there any *organizational givens*, things we cannot change?
- What are the *requirements* for the process?
- What are the *principles* for the process?

To answer these we must question the relevant stakeholders: the process owners, managers, customers and sponsors. How we ask them will be a matter of choice: a workshop, a round-table meeting or separate interviews, for instance. However we go about it, we must document the ground rules we discover and get agreement – formal or informal as the occasion demands – from those who matter.

Are there organizational givens?

Are there any bounds to our ability to change things? For example:

- **Are we constrained by the current organizational structure?**
  Must the process work within the current organizational structure as it is, with the current roles and responsibilities? Are existing roles and responsibilities up for grabs? Are some roles fixed and others open to change? Can we create new organizational units, posts and job titles?
- **Does our IS or IT infrastructure impose constraints?**
  Are we restricted in what we can expect in the way of person-to-person communications, support for workflow and job scheduling, availability of information, or production and distribution of documents in all their forms?

- **What constraints do the regulators place on us?**
  Are there specific requirements for reporting from our processes? Are there required levels of transparency? Which areas of activity must be open to independent audit? Are there prescribed interactions with the regulatory body? Are there things that must be approved or licensed before we can proceed at certain points?

- **Are there safety, security, or risk issues to be addressed?**
  Do we need independent roles with responsibility in these areas? What interactions are prescribed for them? Are there prescribed procedures that we must incorporate in our own processes? Are there prescribed processes with which we must interact?

- **Are there cultural norms that must be observed?**
  Are we an empowered or a controlled organization? What behaviours do we expect from our staff and our managers that might steer us towards certain styles of interaction or relationship?

What are the process requirements?

The process architecture proves itself useful again at this point. Simply by chunking the organizational activity the way it does, it gives us a basic level of chunking of the requirements on the organization across the processes. If a process is called **Handle a Regulatory Request, Produce a Product Batch, Handle a Customer, Manage the flow of Products** or **Manage the flow of Orders**, we already have a clear focus. Modelling methods that allow arbitrary decomposition leave us with arbitrary chunks and no clear focus.

Pupil:	I hesitate to suggest this, but isn't this an ideal place to think in terms of inputs and outputs? Can't we simply express our requirements for the process in terms of what goes in and what goes out?
Tutor:	Firstly, let's remind ourselves that we are talking about designing CPs and CMPs. A neat end-to-end process like a CP might suggest an input-output approach. But you must admit that it is less likely to make sense for a multithreaded thing like a CMP?
Pupil:	Well, possibly. I'll suspend my disbelief for a moment – I'm happy to concentrate first on the CP – the argument will be stronger there.
Tutor:	OK. Let's take Handle a Regulatory Request as an example. It's pretty clear what the input is. What about Produce a Product Batch? We know from its name what the output is. If you remember, we based our naming convention that way. But I can imagine a whole pile of other things that we shall require of a process that we would be hard-pressed to express in terms of inputs or outputs. Here's a few to be getting on with: *Customer has been notified that their complaint has been rejected; Management have been informed of price; Batch is in regulated store correctly labelled; Marketing are satisfied with the timing.*
Pupil:	I see. It's often not just the thing we want to identify – complaint rejection, price information, management, batch, and so on – but its state.

Tutor:	Exactly. By working with final states we can be more general, and we're not forced to contort everything into the straitjacket of outputs. We can make the same case for working with initial states instead of inputs. There is another danger we need to be careful of: that of writing process design when we should be writing process requirements. For instance, if we write 'The customer must receive a letter of apology,' we might have made a design decision: the apology will take the form of a letter.
Pupil:	You're saying in effect that we need to work with an abstract model?
Tutor:	Yes, I am.

The requirements on a process

When we draw up the requirements for a process, we shall be looking at it from several angles:

- The starting state.
  What will be the state of affairs when the process starts? There might be several possible starting places – triggers – and we shall need to cover each; this is especially true of a CMP. We can think in terms of how we would write the state description at the beginning of the thread on the RAD, whether for a CP or a CMP:

  - 'Approved plan in hand.'
  - 'Customer call is switched to the appropriate desk.'
  - 'Specification for batch has been approved but resources not yet allocated.'
  - 'Project definition available in Project Inception Document.'
  - 'A request has been received for a new clinical trial supply to be made.'
  - 'A project has reported an overrun in budget.'

- Any actions required in the process.
  Perhaps certain checks are required or certain quality controls must be carried out:

  - 'All intermediate products must be tested for specified purity.'
  - 'The unit must be shown to have achieved greater than 99.9 per cent reliability.'
  - 'All materials leaving the plant must be appropriately labelled.'

- Any business rules governing the process.
  Perhaps certain financial oversight is required, or certain people must be involved in certain decisions or actions. The relationship between a CP and its CMP will appear here. And some rules are what we earlier called 'steady-state goals'. For instance:

  - 'Product Quality Assurance must approve all test plans.'
  - 'Divisional Management must be kept apprised of progress.'
  - 'Resources will be allocated by the Production Planning Team.'
  - 'Marketing need to be satisfied with the timing.'

- 'All pre-customer checks must be carried out by an independent testing group.'
- 'Any Type 5 expenditure must be approved by the Divisional Director.'
- 'All materials supporting the decision are to be retained.'
- 'The Customer must be kept up to date at major milestones in the progress of their application.'

- The required outcomes of the process.

  For a CMP we can ask what the possible outcomes are for each of the identified triggers. For a CP we can ask what state we want things to be in when the case has been handled, i.e. the process has completed. We might recognize several alternatives: success or failure, approved or not approved or returned for rework, and so on. We might describe those alternative states in terms of things that have been produced or changed – but only if those things are necessary and of the essence. We can think in terms of how we would write the state description somewhere on the RAD:

  - 'Approved plan in hand, management informed of content, and projects database updated accordingly.'
  - 'Customer complaint rejected.'
  - 'Batch of chemical in regulated store correctly labelled.'
  - 'Customer has agreed that their call has been sufficiently answered.'

What are the process principles?

The requirements we were looking at in the preceding section were quite functional: the process must *do* this, that, and the other. But there can also be requirements on the process that are more about what the process *should be like*, how it should feel to use, how it should feel culturally, what organizational norms it must conform to, and so on. These sorts of requirements are also typical of an improvement situation: we know what the process should do – we now want to change its feel.

Let's take an example. A large organization with many thousands of computer users had a large infrastructure services group, itself numbering hundreds of staff. Running a company's networks and computers is a thankless task at the best of times: people are rarely satisfied with the PC on the desk in front of them and the facilities behind it. This group was going through a particularly bad patch and relationships with the user community had reached a low. The perception was that the constant battering from the user community had left the group with a defensiveness that was not in the business's interest, but that had become built into its processes. If the relationship was to be improved, the processes had to be redesigned to meet the expectations of customers better.

Getting to the root problem of the perception was hard. But a long session with a group of users, using brainstorming and affinity diagramming, allowed us to reduce the many gripes to a handful of failed expectations about the way that service request calls to the group were handled. Turning these around, we were able to summarize a set of process principles that would lead to more agreeable experiences for users and that should therefore govern the process redesign. When you read them, you will find them very obvious – but then these things always are:

- When I make a service call, don't bounce me from person to person.
- When we discuss my requirements, set my expectations at a reasonable level.
- Don't say something to please me and, in the event, let me down again.
- Be honest with me about timescales.
- Get back to me if you say you will.

These straightforward principles captured what was lacking in the processes as they had ended up. We put them on the wall at the subsequent process redesign sessions to guide the design ... but more of that in a moment.

It need not be the customers of the process that want to see improvement – we would expect management to look for it too, of course. As another example, let me quote a large software development group inside another multinational. A cultural move was under way to increase responsibility-taking and proactive problem-solving, and to add a measure of risk-taking. Existing processes were being redesigned. A workshop of stakeholding managers, from within the development group and its customers in the business, produced a set of principles that reflected the changes that were sought:

- Commitments and commitment points are clear.
- Deliverables are clear.
- Responsibilities are clear (including those for the quality of all deliverables).
- The process acknowledges and attacks risk and makes risk visible.
- There are clear and appropriate financial controls and responsibilities.
- There is clear, fair and unambiguous reporting.
- Everything is change managed.
- It's a collaborative process.
- Quality is central to our approach.

As a third example, I'll say a little about a project with a major charity that was changing its administrative processes to take advantage of a new software system being installed. The new system made possible more streamlined processes, so the processes and system were designed in parallel. The *functional* requirements for the system emerged quite naturally as the design proceeded. As part of the preparatory work, the more abstract requirements of the system and the process were identified in sessions with both users (the administrative staff) and management. They covered the need for *confidentiality* about cases (the system administered grants to individuals in need), *speed* (grantees typically had low life expectancy so a protracted grant approval process could prove useless), and *sensitivity* (ensuring that, for instance, the process responded appropriately if the intended grantee died before approval).

Process principles can – of course – conflict. If we have visited the different stakeholders and asked them for their principles, we may easily end up with conflicting pressures.

However we acquire them, these process principles are important enough to be written in large red letters on sheets that are on the walls when the process design workshop is run. Both during and after the workshop, it's our responsibility – as workshop leaders – to be constantly checking that the design is in line with the principles. The principle 'When I make a service call, don't bounce me from person to person' had clear implications for the way that the roles involved interacted: the person picking up the original call would not be able to relinquish responsibility for it by passing it off to someone else; they would have to remain in the loop and in control. 'Get back to me if you say you will' meant having reliable prompts on the supporting call management system which were acted on. An estimate of time-to-fix would be given to a caller only after enough investigative work had been done and the estimate had been checked: wild guesses were not enough. If an estimate for completion was likely to be too unreliable then the response would be staged and an estimate given for completion of just the first stage, which might be a visit to the caller's desk, for instance.

KEY POINTS

The process architecture tells us what processes must be designed for a new (part of an) organization.

Before we start the design of a process we must:

- list the organizational givens, the constraints on our design freedom;
- list the principles that are to govern the design;
- list the requirements that the process must fulfil, reconciling conflicts where they arise between different stakeholders in the process.

Requirements for the process should be expressed in terms of:

- its starting state(s);
- any activities that must take place;
- any business rules governing it;
- the required outcome(s).

THE PROCESS DESIGN PROCESS

We are designing a process from scratch. We know from the process architecture which process we are designing. We have established the constraints, principles, and requirements that apply. It's time to run a design workshop.

Preparing for the process workshop

The process architecture tells us the basic boundary of the process:

- If it is a CP, the initial state will be some variation on *Case in hand* and there will be one or more potential outcomes, presumably at least one of which will correspond to 'success'.
- If it is a CMP, we have a list of candidate triggers from Chapter 5. Each of these will have one or more appropriate outcomes.
- If it is a CP, we know which UOWs are in turn generated and hence which CMPs must be approached to get cases of those UOWs under way.

As there is no existing process, we do not do any prior briefing. It will be enough to get the ground rules in place as described above. We can now run a process workshop along the lines of the generic one described in Chapter 8 – but with some major adjustments for this greenfield situation. The sequence of steps will vary from situation to situation but the following will give some structure to the event:

- From the outset we have flipchart sheets on the wall with all the key inputs:
  - the (relevant part of the) process architecture;
  - the name of the process;
  - a list of candidate triggers;
  - one or more potential outcomes;
  - relationships with other processes;
  - the organizational givens;
  - the process requirements;
  - the process principles.
- Checking the list of organizational givens on the wall, rather than brainstorming existing roles, we brainstorm the necessary areas of responsibility – abstract roles – corresponding to:
  - mandatory or specialized skill sets;
  - mandatory approval;
  - mandatory oversight;
  - mandatory review;
  - restrictions imposed by IT;
  - any activities that must be carried out independently.

- Checking the list of process principles on the wall, we brainstorm further areas of responsibility – abstract roles – corresponding to:
  - desired skill sets;
  - desired approval;
  - desired oversight;
  - desired review.

- We brainstorm basic actions within the process, allocating them to abstract roles and noting any ordering that is immediately apparent.

- For a CP, we brainstorm the sorts of interactions we expect the process to have with its CMP.

- We identify business-derived decisions and alternatives.

- We gradually lace together the abstract roles with interactions that support the collaborative content. Our first model will be an abstract model.

- We use the process principles to validate the design.

- We check that all the requirements have been dealt with.

- We check that the organizational givens have not been contradicted.

- If appropriate, we then decide how we want to implement the process in terms of concrete roles, concrete mechanisms for interactions, and so on. In going from the abstract to the concrete – i.e. as we cut up the pizza – we shall find ourselves creating new interactions between the concrete roles.

- We identify how existing or potential information technology solutions could turn abstract things into concrete. (Tutor: That sounds dangerous, doesn't it?)

Finally, there is a thread that we left dangling in Chapter 2 and which it is now time to tie off. Suppose we are designing for a steady-state goal such as 'Bank balance is always positive'. We must ask 'What might change the bank balance?' and ensure that only actions that keep the bank balance positive are possible. This might mean, for instance, only allowing spending if a prior check has shown that there are sufficient funds to meet the cost.

Slightly trickier, we might require that 'Marketing are always aware of the status of new product developments'. We perhaps recognize that there will be periods when Marketing haven't heard some detail of a new development and that the best we can do is to minimize those periods. Are there actions or interactions in place to restore that part of the state after process elements that might have changed it? Where an action can leave Marketing out of date, we can ensure there is an immediate interaction to bring them up to date.

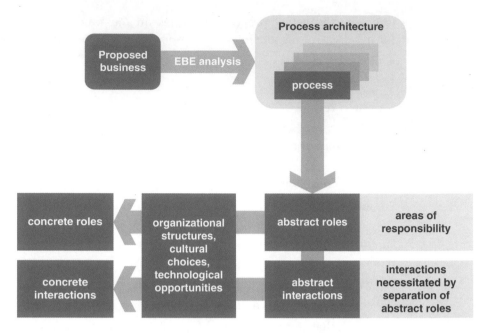

FIGURE 11.1 *The general scheme for process design*

SUMMARY

Pupil: You haven't exactly provided a recipe for designing a process.

Tutor: No, and I don't think it's possible to write one. But we have listed what we need to do to generate the information we need to do a design. Design is not a mechanical process – it is all about invention and thinking – it's above all a creative process. The best we can hope to do is to identify what will help us get everything on the table before we start, and check that what we create meets the requirements.

Pupil: I'm also nervous about the fact that we haven't touched on issues such as whether or not the design is acceptable to management, whether it can really be supported by the IT infrastructure, whether it would actually work, whether it would work fast enough.

Tutor: You're right to be concerned about these things, but we did say that all those process *programme* issues are for other books and other tutors ...

12 Processes and information systems

Covers the role of the approach in constructing an IS strategy for an organisation, and in the design of traditional information systems.

INTRODUCTION

We know that for every UOW we are likely to find, in some measure, three processes: the CP, CMP and CSP. Each of these can be expected to be both a provider and a user of information, and hence a potential user of computer-based information systems.

The case process

IS support for the CP will be about the capture, storage, presentation and processing of information about individual cases. When I call my water company, my name and account number (and perhaps the number of the telephone I am calling from) will allow the information systems to bring up my details and a summary of recent contacts on a screen in front of the person taking my call. The time frame of IS support for the 'lifetime' of a customer contact will be short- to medium-term: some contacts will be cleared during the call, some will need action that will continue for perhaps weeks as visits to the customer's premises are arranged, take place and get closed off. The time frame of IS support for the 'lifetime' of a customer will be medium-term to long-term, since an instance of the CP for an individual customer will last as long as the customer is a customer.

The case management process

IS support for the CMP is about the flow of cases at any one time – typically rates and levels – and about exceptions that are raised by the CP instances. As Call Centre Manager, I expect to see on a screen in front of me who is doing what, what the waiting queue of callers looks like, and how it is building or falling. The actual allocation of calls to call centre staff is being dealt with automatically by my smart telephony system, leaving me to manage the staffing levels and deal with exception conditions. The time frame for such IS support will be about minute-to-minute management of the centre. In other cases the time frame could be years.

The case strategy process

IS support for the CSP is about the longer-term capture, storage, analysis and presentation of past cases. We will think in terms of data warehousing and data marts, the ability to slice and dice the case data, computer simulation of the business and its flows, exploration of trends, experimentation with different future scenarios, and modelling of new structures and processes. The time frame for such IS support will be medium and possibly long term, depending on the time horizon of the business.

RIVA AND TRADITIONAL INFORMATION SYSTEM DEVELOPMENT

From Riva process architecture to IS strategy

Utility company R sought a business-driven IS strategy, one driven by the needs of the business rather than the capabilities of technology. For any business, the issues that drive management action vary in time as priorities change. Those business drivers will have different time horizons and the role of IS is to support them accordingly. But the translation from business drivers to information technology strategy is a big one: we needed to break it into a number of smaller steps that could be taken individually with senior managers.

The Riva approach in this area is summarized in Figure 12.1. The first step was to brainstorm those business drivers: ignoring information systems, what's driving the business and changes in the business today and in the foreseeable future? We developed a hierarchy of drivers with lower levels making the 'woolly' upper levels more specific and finally, where possible, measurable. At the highest level in our case study the drivers included safety, expanding the customer base, and satisfying the regulator. Each of these was then decomposed in turn to bring the area into finer focus. A driver such as 'We must expand the customer base' might be decomposed into drivers such as 'We must have more flexible pricing,' 'We must package our offerings better,' 'We must improve the quality of our *A* service,' and 'We must understand better who is buying what.' The driver 'We must have more flexible pricing' became 'We must know which of our assets are being used and how much' and 'We must be able to construct more flexible contracts.' This sort of analysis might already have been done by the organization and we can pick it up directly; otherwise, a traditional facilitated workshop will yield the information.

While this work was progressing, a simple Riva process architecture was constructed for the organization.

We worked with senior managers responsible for large chunks of the company. They examined the list of processes in the process architecture to see which potentially had the most to gain from having accessible information. Simple round-the-table scoring allowed us to rank the processes according to their 'information potential' – clearly processes

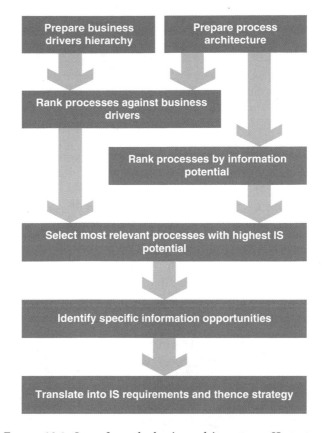

FIGURE 12.1 *Steps from the business drivers to an IS strategy*

with the most to gain from strong IS provision would be of greatest interest to us in developing the IS strategy. They examined the processes to assess how far each could help the organization meet the demands of each of the 'leaf-node' drivers in the driver hierarchy. A traditional matrix of drivers against processes was used to score each process according to its contribution to each driver and thence to derive a further ranking. By simply multiplying the process–driver matrix by the information potential rankings vector, we were able to quickly prepare a shortlist of the processes that were most important to the business *and* had most to gain from strong IS provision. In a final bout of workshops, we drew out improvements in information quality or accessibility for each of those processes, knowing now that we were addressing the big opportunities. These requirements could then be translated into requirements on the IS provision.

At this stage, we were of course simply characterizing the IS provision in terms of what it would *achieve* as opposed to what it would look like: 'Timely analysis of the usage that each asset component had and of its costs, available at the desk of group *D*,' 'Timely costing of alternatives packages, available at the desk of group *E*,' and the like. The remaining step was to decide the technologies that could be used: data warehousing, client-server architectures, web-based information gathering etc. This was a purely technical judgement to be made by those familiar with the technologies concerned, but the senior managers now had a clear vision of why those technologies were relevant to their business – one of the hardest steps to make – and we made the bridge for them using the process architecture. If, say, a particular IS architecture was proposed to satisfy the information needs of the asset CSP, managers could trace the decision back to the business drivers concerned with their need for pricing flexibility, and the proposal became meaningful.

From Riva process architecture to IS gap analysis

Coming at the question from the other direction, the IS group at utility company W wanted to develop an IS strategy that was aligned to the business in a way that was accessible by senior management. The strategy would need to map a path from the existing provision to the new. Again, the process architecture provided the bridge.

Figure 12.2 summarizes the approach.

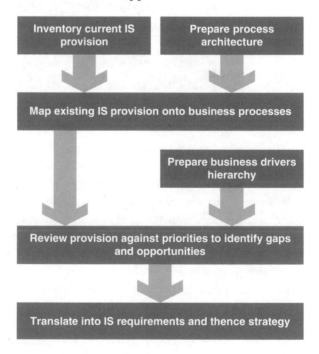

FIGURE 12.2 *Steps from the current IS provision to an IS strategy*

For an IS group, the preparation of a process architecture can be kick-started by using the corporate data model as a checklist of UOWs. (There will be many things that an organization collects data about that do not represent the 'work items' that we look for in UOWs, but it would be strange for there to be a UOW about which no data is kept.) While this work was going on at W, a parallel team was preparing an inventory of existing systems and their users. Those systems were then mapped onto the process architecture – knowing the users of the system allowed us to determine what processes they were involved in when they made use of the system.

The result was a gap analysis. Certain processes were well supported by the current IS provision. But an analysis of the business drivers of W pointed to process areas of the business that would become important before long and which would need increased support from IS. It was now possible to prepare a 'shopping list' of systems and architectural changes necessary to make them possible. The business case made to senior management could now be aligned to the business's processes and drivers, offering the opportunity of a much richer form of justification than a cost-benefit comparison – the IS group could be painted much more easily as an *enabler* for the business, a case that has traditionally been difficult for any IS group.

Figure 12.3 summarizes the approach to using a Riva process architecture in preparing an IS gap analysis.

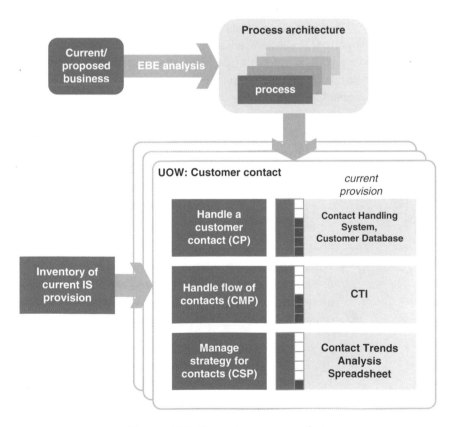

FIGURE 12.3 *Preparing a gap analysis*

Preparing an appropriate process architecture

Using a process architecture in these two ways places few new requirements on the way we build the process architecture in the first place. The approach in Chapter 6 requires little adjustment.

The one decision we must make is whether to include in the process architecture the processes for designed as well as essential UOWs. An IS strategy that is aligned to the business strategy will, minimally, provide an adequate level of support to each of the processes for each of the essential UOWs. But an adequate IS strategy will also support the way the organization has chosen to do its business – in other words the designed UOWs – so we can expect that these will appear in the analysis.

Using a Riva process model in traditional IS development

Because Riva gets to the heart of what is going on in an organization, the process models it produces are ideal starting points for the development of traditional information systems. Each such development should start with some sort of analysis of the business and, thereby, where and how an information system can help. But traditional approaches have not looked seriously at the *process*, something that we now have the machinery to do.

Traditional methods (such as SSADM) concentrate on the information needs of the individual and the way that information finds its way between individuals and groups through 'plumbing' between them and some database. To do that they use techniques such as *entity relationship modelling*, *data flow modelling* and *entity life history modelling*. The information collected during a Riva analysis connects tidily into these:

- When we look at the UOWs in order to get at the case structure of the organization, and model their relationships as a step to understanding the likely relationships between processes, we have a sound starting point for preparing the traditional entity relationship, or data, model. The latter will add more entities and more (static) relationships.

- By examining the actions, interactions and decisions within roles, we identify the information needs of those operating the process.

- When we examine the information inputs and outputs of actions and the flow of grams in interactions, we build the basis for a data flow model of the process.

- When we follow the history of a UOW entity (typically through the CP that deals with it), we can map out its state history and hence have the basis for its ELH.

  - The flow of an entity through the actions and interactions will be shown on the RAD; the actions and interactions through which it flows will change its state.

- The state changes caused by the process will be shown on the ELH. We can check where in the process those state changes occur and ensure that the ELH and the RAD tell a consistent story.
- The attributes of the entity, as listed in the ER model, should cover its state. We can check that the attributes are sufficient to describe the entity's different states.

In short, having done its job in helping us understand, analyse, or re-design a process, a RAD also provides us with a sound starting point for the data-oriented development of an information system to support that process.

Information systems, and ERP systems perhaps even more so, are notorious for being wet concrete before they are built and set concrete once they are built. Worse, they set in concrete the very processes they support, making them hard if not impossible to change once the concrete of software has set. All too easily, crazy processes become ossified, a phenomenon known as 'paving cow paths': things have always wandered this way in the past and now, thanks to the information system, they are doomed to wander this way in future. In the very worst case, it is also a matter of 'paving cow pats'.

The message here is, of course, that we should take an abstract perspective in our RAD in order to ensure that our new process support system does not simply mimic, say, the existing paper-based system but instead takes advantage of the potential of a fully electronic environment: document management, document imaging, smart telephony, web services, PDAs, network communications, group support products etc. Our process models will therefore be strongly abstract.

RIVA AND OBJECT-ORIENTED SYSTEM DEVELOPMENT

One of the central notions in Riva, whose roots go back to the early 1980s to the formally-defined object-oriented language SPML, is that of the *type* (or *class*) and the *instance*. We saw how everything on a RAD is actually a type, and that we can 'run' a RAD by looking at how, when and what instances of types of roles, actions and interactions are created as the process runs. Riva's object orientation makes it a natural business analysis precursor to the use of object-oriented software development approaches and in particular, the object-oriented notations in UML (Unified Modeling Language). Riva's two notations are not a replacement for those of UML any more than UML's fifteen or so are for Riva's. Their shared object-oriented roots make them good bedfellows but some apparent similarities must be handled with care. Unfortunately, the attempt to crowbar UML into the world of business process modelling has been made and we must hand over to the Tutor to rebuff it ...

During a break, next to the water cooler

Pupil: I've had a lot of dealings with software developers recently and they have really taken to UML for software specification and design. I know that a number of people also use UML in business process modelling. It sounds like a sensible idea – so why try and persuade them to use a different approach for the business process side from the software design side?

Tutor: Well, you've given me the reason without realizing it: UML and all its diagrams come from the world of software engineering. Let me tell you a story. You might have come across the IDEF0 notation. Its precursor (SADT) was invented in the 1970s at TRW for the analysis and design of computer systems using 70s technology, for which it was well suited. It has been terribly abused by being transplanted from the world of system design to the world of business process modelling, a purpose it was never intended for when it was invented. Like UML, IDEF0 has no business- or organization-oriented semantics. Worse, it was based on hierarchical decomposition which made great sense in the 1970s when software development had that paradigm, but it makes no sense in the modelling of organic business processes.

We shouldn't be surprised that using software-oriented methods to model anything other than software systems proves hard work. It's possible, of course, but then we could model business processes with Turing machines ... and much good would it do us. My message is that Riva's diagrams are there for a purpose, and UML's for other purposes. Attempts have been made – in particular by tools vendors eyeing up the BPM marketplace – to adapt UML for business modelling despite its lack of business-oriented semantics.

Pupil: You're showing your age ... But let's take UML's use case diagrams – they're very popular and they've been proposed as a starting point for building a model of a business – I understand *The Object Advantage* (Jacobson *et al.*, 1994) was the influential first attempt. What's wrong with them?

Tutor: Well, let's check on definitions first. A use case was originally defined in the world of computer systems as a sequence of transactions designed to produce value to a user of the system. If we translate that out of the world of computer systems to the world of business, we get something like 'a sequence of transactions designed to produce value to someone outside the business'. Fine, except that there's little agreement after more than a decade on what use cases are and how to work with them – what does that tell us? The real meat in use cases comes with the *use case description*, which is typically a page or two of text that describes the use case. So what's happened? We've ended up writing serial text to describe a definitely non-serial thing. All sorts of contrivances are then necessary to handle exceptions, alternative routes etc. These are just attempts to render a complex network structure into serial form: text! The question of use case 'extension', for example, confuses the issue and, it appears, practitioners. At best, use case practitioners apologetically suggest that use cases are just an exploratory tool. Not much good for real understanding then! As a way of finding out what processes an organization has, what its dynamic structure is, use cases prove very weak.

Pupil: OK, accepting that a set of use cases would only 'explore' the business activity as you put it, what about the other notations that are then used to support the use cases: the state diagrams, activity diagrams, interaction diagrams, and so on?

Tutor: I think you've answered your own question again: we are now going to collect a mass of detail into a number of quite separate and very *technical* viewpoints: it's down to the reader to try and integrate all these different viewpoints into a cogent picture of what actually happens in business terms in the organization. So the structure of the resulting 'model' isn't business-orientated: it's concept-orientated. And the concepts – *interface objects*, *control objects, entity objects* – are not business-oriented concepts, they are technical concepts.

People who have used UML diagram notations for business process modelling often remark on the fact that several diagrams with different notations are needed to capture a single business process. As a result, (i) it's hard for an analyst – let alone someone in the business – to easily see from the multi-concept, multi-viewpoint set of diagrams what's really going on out there in the organization, and (ii) the many and detailed cross-relationships that arise between the diagrams have to be maintained, and this makes for problems with change as well as understanding.

Pupil: So you maintain that Riva's concepts such as the UOW, the case process, the case management process, creating a responsibility, responsibility roles, organization roles, actions, interactions, decisions … that these are the proper metaphor for business process modelling?

Tutor: Of course. They have analogues in the real world. Real-world people recognize them. Moreover, all of them are underpinned with an object-oriented theory with formal semantics, but we don't expose the poor business folk to that. In this book – for analysts – we have used the object-oriented concepts and even the word 'instantiation' but, as I stressed at the outset, we don't need to trouble the business folk with them.

The problem is that with a UML toolset on our PC, it's impossible not to slide into ways of thinking that are perfectly valid when thinking about software design, but that make no sense when thinking about the real world. A UML business modeller would tell us that in a restaurant we have an *Order* object which must have an associated operation *Which dish*, which is stimulated by, for instance, the *Order Handler* interface object – I have read this somewhere. This would feel perfectly natural if we were discussing a software design, but 'invoking the operation *Which dish* from an *Order* object' has absolutely no analogue in the real world. It means nothing in the real world. To claim that it models something that happens in the real world is nonsense. It might be a sensible description of some class interactions in a software system, but it has no meaning for someone in the restaurant business.

Pupil: Isn't this just a question of terminology?

Tutor:	NO! It's a question of having the right metaphor, the right idiom. The metaphor of 'objects responding to received messages' is fine where we are building a software system using object-oriented tools that share that metaphor. But that metaphor has no analogue in the real world: chefs do not send messages to orders to find out what dishes are required.
Pupil:	Point taken.
Tutor:	Don't despair. By all means use UML where it was intended to be used and is entirely appropriate: in the analysis and design of software. So, let's move on and look at how we can make the transition from a sound *business* model in the form of a Riva process architecture and a set of RADs to a *technical* description of a system in the various notations in UML.

From Riva PAD to UML use case diagram

We know that a Riva process architecture is precisely based on a clear characterization of the organization's business in terms of its EBEs. A clearly defined analysis path leads us to identify all the organization's processes and produce a second-cut process architecture. CPs in particular can be identified with business use cases: after all, each CP delivers some value to a 'user'. A CMP can also be identified with a 'large' business use case, one that we will be able to decompose when we examine the different threads within it at RAD level.

A traditional problem with use cases is knowing at what 'level' we should be thinking. Are all the following system use cases: 'I go to hospital,' 'I visit the Haematology Department,' 'I have my blood pressure taken,' 'The nurse records my blood pressure'? This problem might not have been solved for UML system use cases, but by using the notion of 'fainter and fainter' UOWs that we introduced in Chapter 6 we are now able to get our heads round the problem, rather than simply moving it from the system arena to the business arena by inventing 'business use cases'. Our Riva analysis will readily identify, in particular, *Patient Hospital Visit*, *Haematology Visit*, *Blood Pressure Reading* as UOWs, and clearly identify their dynamic ('generates') relationships. Because the process architecture was produced without consideration of technology or systems, we can be confident that the corresponding business use cases are devoid of system design – a good thing.

From Riva UOW diagram to UML object model

UML class diagrams must not be confused with Riva UOW diagrams. The former show all relationships and are avowedly static views of the world. The latter show only dynamic relationships, as they are a step towards the all-important *process* architecture. We can certainly expect that essential and designed UOWs will appear on a UML class diagram for a system supporting the organization characterized by those UOWs. But neither is a replacement for the other. That said, a Riva UOW diagram is a perfect starting point for a UML class diagram, being based on a strong analysis as described in Chapter 6. To the dynamic relationships in the UOW diagram we must add the static relationships of interest in system development.

From Riva Role Activity Diagrams to UML activity diagrams

At first glance, UML activity diagrams with swim-lanes suggest Riva Role Activity Diagrams. But they lack a number of important concepts. Firstly, a swim-lane is a static thing, the responsibilities of a certain class, rather than a responsibility that can itself be instantiated or even passed around – a swim-lane cannot generate the network of collaboration that characterizes organizational activity. Secondly, where activity flow crosses from one swim-lane to another, it is just that: a movement of the locus of activity; the implied interaction is not recognized, the contribution made by each end is not recognized, and the object-oriented integrity of the interaction is lost. Roles don't just partition activity, the way that swim-lanes do: they also express the interactions – the strands of mozzarella – that are generated by the partitioning. The activity diagram ignores interactions as joint activities that synchronize state, which is what happens in reality. A UML activity diagram, taken together with an interaction diagram and sequence diagram, could cover the ground of a RAD, but without the clarity or precision. For instance, UML interaction diagrams capture some of the dynamics of a 'system' in terms of messages passed between objects, though typically they are potentially incomplete since 'returns' are not always drawn, for instance, being regarded as 'clutter'.

Preparing the process model for object-oriented system development

If we are starting with a clean sheet of paper, our process model should be of the abstract variety so that subsequent work is not tainted with design decisions that should be left open at this early stage.

RIVA AND WORKFLOW SYSTEM DEVELOPMENT

A Riva process architecture, complete with processes for designed UOWs, exposes the workflows to be supported by a WFM system. Such systems are typically designed to support CPs in which a thread of activity leads from case inception to one of a number of possible outcomes. A degree of branching of the flow is generally supported and some parallel working, though this is typically less general than we would like. Take the RAD in Figure 12.4 as an example. It shows a simple, single-threaded CP for handling a job application. It is the sort of RAD that might result from work to analyse and design a process that we want to support through workflow technology. We have taken it more or less to the point where we can start to decide which areas of the process can be supported with which sort of technology.

Every applicant gets an acknowledgement in the form of a standard letter (a P11); some are turned down on the basis of the application form

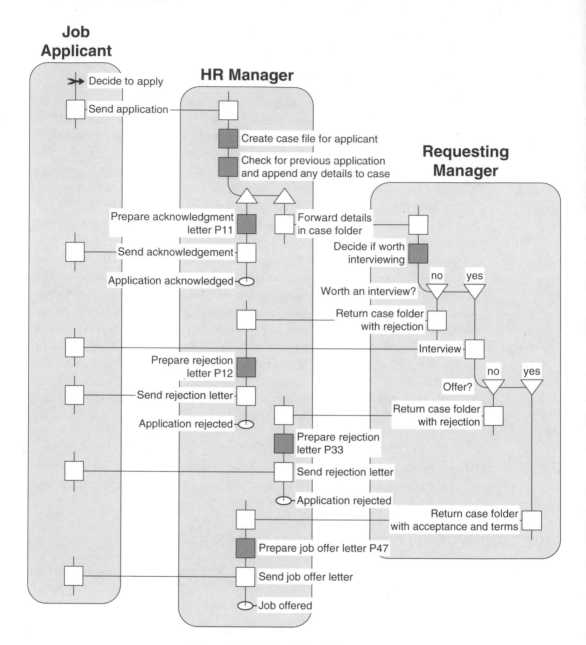

FIGURE 12.4 *A simple workflow process*

and receive a P12 letter; others get an interview but are turned down with a P33 letter; and successful applicants get a job offer in a P47 letter. The *HR Manager* handles all interactions with the *Job Applicant* except for the interview, which is done by the *Requesting Manager*. Let's examine how workflow technology would naturally support this sort of process.

When the *HR Manager* receives an application from the *Job Applicant*, they put together a 'case file' for this application. In a manual system, that file could be just a labelled manila envelope which is then passed by hand from person to person. In a 'pure' IT system it could be a new record on a central database that can be accessed by anyone involved in handling the case. In a WFM system it would be an electronic case folder that the system conveys with the case on its travels round the workflow.

Before processing the application further, the *HR Manager* checks whether the *Job Applicant* has ever applied for a job with the company before and, if they have, attaches the details of the previous application to the case folder. It might well be that there is an existing database which holds such details and which therefore the *HR Manager* needs to interrogate through an existing legacy system. We can therefore expect them to use some standard forms to get at that database and to transfer the results into the e-folder for the case.

The e-folder can now be forwarded to the *Requesting Manager* for further work. That interaction is a 'hard' interaction and could be simply supported within the WFM part of the system.

Subsequently, the *HR Manager* will interact with the *Job Applicant* in various ways: acknowledging the application, turning down the application, getting in for interview, turning down after interview, and making a job offer. Each of these involves preparing a standard letter. Here is an opportunity for traditional personal productivity tools: in particular a word processor, perhaps used in conjunction with the e-folder to get applicant-related details automatically into the letter. In fact, we can imagine an implementation of this system which does not require the intervention of the actual *HR Manager* at all: the system could prepare and despatch the letters automatically on their behalf.

What have we seen with this simple example?

- Some actions require the individual to have access to the right information at the right time and perhaps to update that information. Such information needs are traditionally supported by 'plumbing' people into corporate or function-related databases – the domain of traditional database products.

- Other actions require the individual to prepare material, perhaps drawing information from a variety of sources. Here we might integrate databases with personal productivity tools such as word processors and spreadsheets.

- Some interactions and workflows are 'hard', that is they are predefined and straightforward in nature, involving the transfer of materials. They are precisely the sort of thing that WFM products are designed to support.

- Other interactions are 'soft'; that is, they cannot be predefined in detail but can still be mediated by workgroup computing products.

- Other interactions will still remain the province of the face-to-face meeting and the phone call.

Preparing a RAD for workflow

Since a RAD is a natural way to capture a workflow, little extra needs to be done when preparing a RAD with a workflow implementation in mind. The only significant issue is the need to be aware of any limitations the WFM system might impose on the complexity possible in the workflow itself.

SUMMARY

KEY POINTS

A process architecture provides a sound basis for ensuring that an IS strategy is aligned to the business, and for gap analysis.

An abstract RAD provides a solid starting point for traditional data-oriented system specification and design.

The object-oriented nature of a RAD makes it a suitable precursor to a detailed implementation analysis using notations within UML.

Provided we take care with process complexity, a RAD is an ideal starting point for direct implementation of a workflow on a WFM system.

13 Processes and process systems

Covers the role of the approach in developing BPM systems in which agile and mobile processes replace static data structures.

PROCESS MOBILITY – THE THEATRE OF THE THIRD WAVE

One very noticeable thing about the email example in Chapter 7 is that, as the world 'runs', so the entire 'email process' evolves into an increasingly complex and intertwined beast. There is a cascade, a flux, a continuous succession of changing process instances that are created and that peter out. Every email corresponds to a process instance. Every thread corresponds to a process instance. Every conversation corresponds to a process instance. And every process instance – whether it is for an email, a thread or a conversation – is associated with an individual with an email address. I can turn that round and say that every email, every thread and every conversation is a process (instance). As a conversation develops and as more conversations are started up, there are an increasing number of process instances running in parallel – and we have increasing concurrency.

Email addresses are the lubricant: the *mobility* of email addresses helps create and cement the link between the process instances (and their associated individuals). By knowing email addresses and passing them around within emails we can expand the web of threads and conversations: the running process evolves. Linkages – in particular between role instances – are created and changed. If I send an email to a group of people and I make their email addresses visible in the message header, I am effectively introducing them all to each other – they can now communicate amongst themselves and the conversation can blossom. In Riva terms, process instances know about each other and can communicate thanks to that knowledge. The process evolves as a constantly mobile network of parallel, interacting and communicating threads of activity.

So, the active process evolves but the definition remains constant. Put crudely, the same RAD will be 'operating' tomorrow as is 'operating' today. Or will it? We need one more level of dynamism: on-the-fly change of the definition. Let's listen in on our Tutor and Pupil.

Overheard at the water cooler

Pupil: A while back we looked at the organization as a theatre – a rather chaotic theatre, which proved to have lots of stages with plays being performed, actors running from role instance to role instance and a mass of interaction between role instances and process instances (performances) going on constantly. But I did notice one thing: the plays were fixed. Is that the only constant in this madness?

Tutor: I'm afraid not. It's about to get much worse. The scripts of plays can be rewritten. We might have liked *Hamlet* the way Shakespeare wrote it, but we might also consider changing it for modern times – strengthening the substance abuse angle at the end for instance.

Pupil: Eeeek! So who is it that is changing the script?

Tutor: That depends. In some situations the original author might come along and produce new versions, perhaps rewriting the final act to give it a different outcome, refining some of the smaller parts for better characterization, or removing unnecessary material. When a new performance of the play is about to start, the actors can use the new version.

Pupil: That's understandable in the real world: we have all sorts of reasons for wanting to change the way we do things – our processes. But presumably the whole business of changing a script happens outside the theatre?

Tutor: NO! The scripts are in the theatre – that's the only place you can change them. Not only are they in the theatre, they are part of the subject matter of the theatre: in other words, you can get hold of them. Now, you can only change a script by using the *process* for changing scripts. Handle a Script – let us say – would be just another process, indeed a CP. Scripts are UOWs! The whole point of this theatre is that it is where *everything* happens – there is no 'outside' – and one of the things that can happen there is that you work with your processes. Putting it another way: the theatre supports you *in* your processes, by managing all the instantiation and concurrency; but it also supports you *with* your processes, by giving you all the means you need to write new ones or change existing ones.

Pupil: I can accept that a script can get changed and new performances use the new script. But presumably any performance in progress is unaffected? Please?

Tutor: Why so? Why should a performance not switch to the new version as soon as it's available? Why should it be forced to carry on with the old one? In some situations that might make sense – for consistency reasons perhaps – but in principle we don't need to make performances stick with old scripts.

Pupil: I'm struggling ... I have a picture of a 'master' script which could be changed. Any performance in progress might continue with the old version or switch to the new version. But at least they are using a master.

Tutor: Who said anything about 'masters'? Why shouldn't a performance use its own variation of a script?

Pupil: ... because ... they ... OK, why not? So they might start with the 'standard' script for Hamlet but decide to change it in some way for just this performance?

Tutor: Of course! They're doing a lunch-time slot, they're ten minutes in and realize that people don't have time for a full *Hamlet* so they quickly do a rewrite and present a reduced version. There is nothing fixed or sacred about a process.

Pupil:	Now I'm getting concerned about the sort of chaos that will ensue if we let everyone tinker with processes as they please.
Tutor:	Then don't let everyone tinker with processes as they please: you must script the Handle a Script process to control what can and cannot happen to a script. Everything is in the theatre, including control over use of the theatre.
Pupil:	My head is hurting.

In Appendix C of his book on the Third Wave (Smith and Fingar, 2002), Howard Smith says 'Any theory of process management must ... break down distinctions between the process of change, the process under change and change in both.' If this makes our brains hurt, we must work at it in the same way that we made the shift in mindset from data orientation to object orientation. Now we must shift our mindset to process orientation.

THE NEW ORDER

In this new environment the business process emerges from the shadows and takes centre stage, not just in our thinking but in our computer systems. In the past, the idea of 'process' was a useful backdrop for thinking about an information system, but the system and its behaviour were expressed in terms of data. With the Third Wave, processes become the *subject matter* of the system, not just the background or underlying framework. BPMSs allow their human users to work *in* their processes and to work *with* their processes.

In this chapter, I want to examine the requirements that this first-class citizenship of the process places on the technology required to build BPMSs and how Riva provides us with the intellectual tools we need in order to use it. Let's reflect for a moment on the history.

I'm old enough to remember when, in the 1970s, a new generation of data-processing systems was touted that end-users would build for themselves using a 'design by example' paradigm. Offhand I can't remember what happened to it, but I suspect the technology of the time wasn't fully up to the job, and end-users couldn't be trusted of course.

Anyway, the relational database folk took over and ensured that control stayed with the IT Department by making the levers so many and convoluted that the end-users had to be kept well away. But at least the organization could have a system that allowed it to differentiate itself, even if it meant depending on the IT Department to deliver it.

There is a reliance in the relational model on the constancy of the business's entity-relationship diagram. Our DBMS can therefore concentrate on adding varying rows to fixed tables; or one could say that it now ossifies our information access system around a fixed set of tables. Of course, if the organization changes how it works the diagram is invalidated, as is our information access system. Suppose we draw a picture of the organization's processes and suppose we base our process enactment system – our BPMS – on it. What happens when the organization or its business changes? Will our BPMS be invalidated? Like our pedestrian information access systems, will it be a drag on the business?

You don't have to be very old to remember how the ERP package folk then commoditized information and process. And how the big consultancy firms were ready with teams of lever-pullers and knob-twiddlers who would 'customize' the standard 'solution' for you. Now you could be like everyone else, more or less, and your staff could move to other employers and immediately feel at home. Process was no longer a business differentiator. But the process was still lodged on the other side of the door to the IT Department.

When BPR was in full swing, by its very nature it called for major step changes in the organization and the way it did business, reinforcing and facilitating those changes by supporting them with ERP packages which brought organization-wide processes that were pre-coded. We had one set of processes one day and a completely different set the next – with all the disruption that went with it. Given that change is the only constant in business, how many organizations are prepared to undergo that sort of upheaval again and again and again, as their business environment changes? We can start to see how important it becomes for the process itself to be there in front of us, the thing that drives the computerized support, and – above all – there for us to change as our business environment changes.

Instead of radical change to the entire organization, suppose we want to start small and support a bit of one process in a part of a department; and that we want to have the end-users – the *process actors* as we could call them – adjust that process until it works the way they want; and suppose that, once they're comfortable, they want to extend its use to the whole department, perhaps incorporating more of that process or even new, connected processes, and then extend out to another department, or a supplier, or a customer. Shouldn't we expect the process actors to be able to do all that? Adapt the process, extend it, spread it, add another etc? Surely we don't want them to have to wait for the Process Technology Department to get round to it. Under the Old Order we had a business–IT divide. Under the New Order process owners design and deploy their own processes, obliterating, not bridging, the business–IT divide. To quote an example given by Howard Smith and Peter Fingar:

> The ideas of the reengineers had created a chasm between thought and action, between process design and process deployment. [The company] began to realize that their problem had never been the capacity to dream up new processes. Instead, the real problem was how to make change happen,

and do so incrementally and on a continuous basis. They needed to make the newly designed processes operational, and to do so step-by-step, process-by-process. Instead of radical reengineering grand designs, what they needed was an actionable and sustainable approach to operational innovation.

(Smith and Fingar, 2004)

Let's imagine another scenario in the New Order. In the world of the fixed and invisible process, buried in the RDBMS or ERP, each patient in our hospital must be processed in the same way if IT have anything to do with it. But surely, when the patient enters the hospital, shouldn't the process that they will follow be customized for them, and won't the health-care professionals then want to adjust that process in the light of the outcomes of treatment? And if new therapies become available, shouldn't it be possible for the patient to be switched to the new process? Not only do we want to change a CP over time, but we may want to customize it for an individual case.

Let's see what all this tells us about the technology.

What is the process equivalent of the entity-relationship model? What is the 'process-relationship model' for the organization? Well, we have seen how Riva provides a fast and reliable way of preparing such a 'process architecture' for an organization, in particular an architecture that stays the same as long as the business stays in the same business. This sort of 'invariant' architecture is exactly what we need: when we choose a process to enact on our BPMS we need to choose it from a constant architecture; when we add a process to our BPMS we must add one from the constant architecture. Our end-users must have a shared view of what the processes are and where their boundaries and relationships are. Such a process architecture provides the bedrock on which we shall 'grow as we go': it chunks the whole process – *the* process (because, in truth, there is only one) – in a way that we can bite off sensible pieces. That's the first feature: our BPMS must be built around a constant process architecture of the sort that Riva gives us.

At any one moment our BPMS will hold – I hesitate to use the term – a *process model*. I hesitate because of course it's not a model in the sense of a model ship, something different from the ship. So instead I'll call it the *process potential*. It's the thing that defines all the potential future behaviour of the process.

Of course, the BPMS will also hold all the past actual behaviour in the form of all past instances: the *process past*. (Indeed, the process past is really part of the process potential: where we have been helps determine where we shall go. Time future is contained in time past.) The BPMS will hold what actually happened, in principle forever: nothing need be forgotten – why should it be? Total persistence. In particular, total *context* persistence. That's the second feature: our BPMS must provide *total persistence*. When I sign on to my PC it just sits there and doesn't do anything. Doesn't it know what I am doing? Clearly not. One of the things I want to do is work on this book. In fact, I am working on several right now, in different ways. I have to remember where in my folder structure I have put files about *this* book and I have to remember what their names are: what *did* I call that file of typesetting instructions from the publisher for this book and where did I put it? Heavens, why am I worrying about the names of things? I just want things to do with this book on my desktop when I decide to work on this book. We only give things names because we need to be able to find them again! I want the BPMS to *know where I am with what I am doing*, and when I ask to pick up the threads (of that role instance) I want to be presented with everything appropriate, everything in context. No more names! ... except for cases: I must be able to tell the system which book I want to work on, but I don't expect to have to know where everything about that book is – it should just come to my desktop. When I choose to work on a particular chapter I must expect to identify it, perhaps by pointing to it in the contents list, but I don't expect to have to know where the files containing it and all its diagrams are: I just want them brought to my desktop again when I point to it. Context is all.

A process is just the way a bunch of people agree to get together – collaborate – to make something happen. The process potential says what, at this moment at least, they've agreed. (The process past is what they have actually done.) If collaboration is what a process is about, then collaboration must be a primitive of the process potential and of the BPMS, the engine that turns potential into past. A Riva-based BPMS would support *roles* and mediate their *interactions* and hence make the intended collaboration happen. That's the third feature: our BPMS must be *collaboration-centric*.

The process architecture will remain constant (if we stay in the same business), but of course we shall want to allow our end-users to adjust their processes ... on-the-fly. I'll put that another way: at any moment an end-user may change the process potential in the BPMS, i.e. change the possible futures. While the process is going on. There's an immediate implication: nothing must exist until it must. The BPMS must not anticipate the future; it must only be in the present – and then only at the last moment. That's the fourth feature: our BPMS must use *lazy instantiation* as its prime mechanism. It must instantiate at the last possible moment. Roles, for instance, unwind action by action, part-interaction by part-interaction: the BPMS does not instantiate all the actions and part-interactions in the role when the role comes into being – it waits until the activating conditions become true and then instantiates.

But can we trust those end-users, the process actors, to change the process? Perhaps, perhaps not. Perhaps the CEO would like to ensure that any change is properly controlled (not the CIO, note, nor even the CPO – the Chief Process Officer). So presumably we shall have a process change control mechanism front-end to the BPMS? Of course not, changing a process is a process and is therefore in the BPMS, part of the process potential. In fact – to be perfectly general – everything is in the process potential. If you can do it, it's in the potential. If it's not in the potential, you can't do it. (And perhaps only the CEO can use the process change process – but that will be in the potential too.) Perhaps different processes have different change processes – these will all be in the potential ... and open to change themselves (I feel a recursion coming on). That's the fifth feature: our BPMS *contains all there is or has been*. Or rather, all there is today – tomorrow we shall change it.

So now we could define a simple process potential for a bit of our department and a bit of our process. We'll use that for a while and tune it, on-the-fly. Then when we're happy, we'll grow the process potential to cover more of our activity, or more of our department, or another department, or one of our suppliers or whatever. We'll grow as we go, the BPMS lazily instantiating and hence forever in step with us, our past always accessible, our future always ours to control.

Processes exist in their own right. There is a process life cycle: discovery, design, deployment, execution, monitoring, control, change. The Third Wave of BPM is more continuous, more incremental than BPR–ERP, and closer to real TQM in that process-centric change is now in the hands of the actors. The process enactment engine is process-neutral; in other words, the process becomes the business of its actors, not the IT department.

In early deployments of the technology we might expect to have to connect legacy systems, packaging them up as web services perhaps. But in truth, that's a dreary prospect. The BPMS would become just glue. A real grow-as-you-go deployment would use the BPMS and the potential *as its entire world* – after all, everything is in the potential and the past: the process and all, everything – who could ask for anything more? (If you could, then put it in the potential!) Because everything is persistent, including relationships, nothing need be named – everything is in context, the end-user's context. We shall have no need of any RDBMS or document management system or ERP to remember stuff. Indeed, they all become redundant – they have always been small, closed worlds and we must do away with them. And we shall have no need of diary systems, workgroup computing (whatever that was), intranets, or any other of the half-baked mechanisms we have invented to coordinate the workings of the organization, to support collaboration. All collaboration is in the potential. And the potential is in the BPMS. And the potential is the BPMS.

Later, next to the water cooler again

Tutor: To calm you down when you started getting worried about uncontrolled changes to scripts, I suggested that you simply needed to get a grip on things using the Handle a Script process. I invite you to think through the implications of the existence of this process.

Pupil: Well, I guess that for each script there is an instance of Handle a Script running. Which means there is a stage where that performance is going on, and that performance has the actual … the paper script on it. Presumably, if someone wants to use the script they get it from that process, from that stage. They can take it away in some form and use it for a new performance on its own stage. So far so good?

Tutor: Yes. But I want to put what you said in a different way: *scripts can be handed around.*

Pupil: Oh dear. One performance can give a script to another performance?

Tutor: Of course. Processes are truly mobile. When an interaction occurs between two process instances, the gram can be a process. In traditional computer systems, data was passed around or messages were passed between objects. Now you can see why I suggested that the object-oriented paradigm was only halfway to full process thinking: the unit of currency is the process. Any thoughts on what would make up the Handle a Script process?

Pupil: Well, there'll be an instance of the process for each script (… and one for itself … oh dear, now I feel a recursion coming on!) … and that instance will have as one of its props the process model – or some representation of the process … as it stands … the current potential. There will be an editing role of some sort and within that role there will be threads that allow the different actions one might want when editing a process model: adding new roles to the process, adding actions and interactions to a role, and so on. It would be nice if all this was done diagrammatically – we would like to edit the RAD itself ideally.

We've talked about versions of things, but since Riva's theoretical under-pinnings are object-oriented I assume that what really happens is that we refine object classes (that define roles, actions etc) rather than replacing them?

Tutor: Yes, that's right.

KEY POINTS

A BPMS holds both the process potential and the process past.

The process potential can be structured using the Riva process architecture.

Individual processes can be represented as RADs.

RADs can be actionable diagrams.

Everything is in the BPMS and in its process context.

If everything is in context, we no longer need to name things except the UOWs.

The BPMS is collaboration-centric.

Change on-the-fly is the norm.

Processes replace data as the new currency.

Revolution is overthrown by evolution: we grow as we go.

Power to the people: a BPMS potentially puts the process back in the hands of the organization.

A final thought. We have talked glibly about 'instances' of processes as a convenient modelling metaphor. Remember that strictly speaking there are no process instances. The clue was in the way that we 'activated' a process: the word 'activated' was carefully chosen. All that actually happened, of course, was that we instantiated the lead role of the process concerned. Just that and only that. We did not instantiate the process. We had no need to. We associate that instance of the lead role with the case it is handling: the lead role has the responsibility for that case and can be associated with it precisely, one-to-one. That case (and its lead role instance) is the closest we get to an instance of the CP. We have to ask what would be meant anyway by 'p is an instance of process P'. At the moment of instantiation of P, what would happen? What properties would p have which were different from those of the case? We clearly could not instantiate anything in P ahead of time – any actions, interactions, other roles – because we do not know now what P will look like when we get to them. Indeed, they might not exist if P is changed on-the-fly. That's what I meant by lazy instantiation.

PROCESS CALCULI

In many industries, business and systems architects strive to create software applications that accurately reflect their business. Sometimes they do not realize that a perfect simulation is their ultimate aim. Architects in other industries know precisely that this is their task. In the logistics industry, companies often model their IT architecture closely around the behavior of the physical logistics networks they monitor and control. ... Gradually [computer system] architects are finding ways to represent the behavior of complex systems — interconnected and inter-related mobile processes – within the business applications they develop. Soon they will realize that mapping business concepts into artificial IT artifacts such as objects, interfaces and procedure calls, should be replaced, or at least complemented, by the process calculus models of the Third Wave. These artificial constructs arose to support the composition of software, not the representation of business. [They] are now looking for methods, tools and systems that are purpose built for business. Increasingly they are looking to business process modeling languages for solutions.

(Smith and Fingar, 2002)

Smith and Fingar hit the nail on the head. Because computers started out as ... computers, things that computed, early computer language development was centred around ways of describing computations on numbers (see FORTRAN). When useful amounts of storage became a realistic matter and symbolic data could be kept *about* things, languages shifted slightly to the side and added ways of describing symbolic as well as numeric data and of 'computing' with data – typically moving it around and rearranging it (see COBOL). Even though our focus of interest has now moved to the *process* about which data must be kept, we still see desperate attempts to use languages for data and data computation to describe mobile processes. It won't work. IDEF0 is one notorious example and UML another. (There is an old joke in software engineering: you can write FORTRAN in any language.) The process-oriented world needs process-oriented languages. The business process world needs business-process-oriented languages. Data-oriented and even object-oriented languages can only be tortured into supporting business process thinking. Process-oriented languages need process-oriented methods – like Riva – to make them work for us.

Calculi for reasoning about systems of processes have been around for decades. Petri Nets have been used for representing systems but suffer two shortcomings: firstly they lack any business-oriented semantics, and secondly they have limited capacity for the sort of cascading and evolution that we know underpins real-life processes. The original RAD notation was based on Petri Nets. It didn't allow processes to change: a process had a fixed structure. The Riva adaptation added replicated part refinement and role instantiation as ways of generating new process at run-time. Riva was based on a derivative of Greenspan's RML that described the dynamics of a process in terms of an object-oriented metamodel with formal semantics. IPSE 2.5 added an operational 'process engine' supporting process mobility.

A crucial question when working with a mass of concurrent activity is how one role instance knows which other role instances it is to interact with. When we walk into a room full of people whom we do not know, we can walk up to someone and introduce ourselves:

Hi, I'm Martyn.
Hello, I'm Angela.

Once this introduction has taken place, we're subsequently able to interact. But how did we know that the thing we walked up to and said 'Hi, I'm Martyn' to was a person? This might sound a stupid question, but suppose you have just joined GlobCorp and your role is to collect project status information from all the project managers – how do you know which of the 1,200 people in the building counts as a project manager? You need to be introduced in some way. When we walk into that room, how helpful it is for the host to take us by the arm and say 'Martyn, I'd like to introduce you to Brian.' Now we know Brian. We might have an interaction with Brian and then Brian might introduce us to someone else with whom we might then interact. Later we might start another interaction with Brian. Sounds like a civilized party.

When we draw a Riva RAD for pure modelling purposes – perhaps as part of a process discovery activity – we don't worry too much about such things. If a Project Manager has to provide a report at the end of the month to their Divisional Director, we simply show an appropriate interaction taking place between *Project Manager* and *Divisional Director*. We don't worry about how, in particular, the instance of *Project Manager* knows which instance of *Divisional Director* to interact with – how they know which is *their* Divisional Director. But if we are going to get a BPMS to run this process, nothing will happen unless (i) a Project Manager can be introduced to the appropriate Divisional Director and (ii) subsequently the two can be correlated when the reporting interaction has to take place. There will be many instances of *Divisional Director* and our instance of *Project Manager* has to be able to say which one it wants to interact with. These notions of introduction and correlation are vital for enactment.

Introduction typically takes one of two forms: *one-to-one* and *directory*. When we were introduced to Brian at the party, it was a one-to-one introduction. We made a mental note of Brian – his face and his name – and Brian made a mental note of us, so that either of us could start a new interaction later. On the other hand, if we are running a small van-hire firm we shall probably have an entry in a trade directory giving our phone number. Someone wanting to interact with us to hire a van can look at the directory listing, choose a hire firm, pick up their phone number and start the interaction with a phone call. At GlobCorp, we might hope to be given a list of all the project managers so that we can interact with them to ask for their reports, without having to be personally introduced to all 42 of them.

Let's look at the Riva equivalents. Suppose I am a role instance, *A*. When I instantiate another role, I clearly know the role instance I have created, *B* – I don't need to be introduced. But *B* might need to know who I am so that we can interact later. So when *B* is created we must tell it who its 'parent' was. Then we can have an interaction where I pass over some terms of reference or whatever. I might also want *B* to interact with a third role instance *C*, so I shall have to introduce the two of them – on a one-to-one basis. I shall have to say to *B* 'You need to know about *C* for this matter' and I shall need to say to *C* 'You need to know about *B* for this matter'. *B* and *C* will then be able to correlate and interact. This introduction takes place over interactions of course ... provided the two ends have already been introduced.

Now, in real life there are situations where we don't have to be introduced for an interaction to occur – what I called *service interactions* in Chapter 2. I don't have to be introduced to the teller at my bank in order to deposit some money. I simply walk up to a teller and they interact with me and I with them. No need for the manager (to whom I have already been introduced) to pop out and say 'Mr Ould, this is Ms Farrier. Ms Farrier, this is Mr Ould.' In Riva terms, some role instances are ready and able to interact anonymously.

Let's stand back for a moment. We know that processes evolve through the instantiation of roles, as new responsibilities are created. In order to collaborate – interact – those newly created role instances need to 'be aware of' each other. Their identities must be available for handing around, for introducing. Their identities must be 'mobile'. It is the mobility of identities that allows the process to change its structure and evolution to occur. As role instances are created and introduced, so the network of introductions develops and the corresponding network of potential interactions can evolve.

The pi-calculus of Milner, Parrow and Walker (Milner, 1999) generalized earlier process calculi by allowing 'channel names' to be dynamically created and communicated, thereby allowing process mobility and a new level of dynamism. Taking its cue from pi-calculus, BPMI.org has published a representation for business processes, the BPML and a Business Process Modeling Notation (BPMN) – see www.bpmi.org. At the time of writing, the situation with BPM standards is still evolving, with a raft of acronyms (BPSS, BPEL, BPELJ, BPEL4WS, ebXML, ebBP...) washing around.

Wherever the ball comes to rest, as long as the metaphor is one of collaborative, concurrent and mobile processes, Riva will help in their design.

SIX VISIONS

Because this process-oriented world is so different from the information-soaked world we currently inhabit, I shall end the chapter and the book with six visions of the New Order. They were originally prepared with David Perrin and Clive Roberts as part of a visioning exercise for a new BPMS. Each is designed to capture one facet of the sort of process-driven world that becomes possible when we base our business systems on our business processes.

Managing globally audited and distributed processes

A Clinical Trials Manager in the pharmaceutical industry can spend a lot of time on planes. Hating to waste time, Francesca uses her four-hour flight back to the UK to work on the three clinical studies she's managing. Via the browser on her PDA, operating over the airborne satellite phone system, she connects to the BPMS back at HQ outside London. After completing security checks, she resumes her role as Manager of the Phase III trial for the new drug Dimoxinol and finds herself taking part in interactions with investigators in France and Belgium. Decisions made and communicated, she moves on to reviewing a proposed change to the protocol – it has been through earlier stages in the process but the BPMS knows that it now needs her approval. As soon as she decides to work on it, the BPMS assembles the necessary documents and places them on her PDA: no need to remember where things are, everything is presented to her in context. A quick scan of the change reveals a serious flaw, so she adds her reasoning and rejects it.

The BPMS pushes that part of the study process on automatically by sending the rejection immediately by email to the change proposer.

The Terathroxine trial in Japan has been suffering from poor patient-recruitment rates – as she finds out when she switches to acting the role of its manager. So what has happened recently? The BPMS memorizes everything that happens, so it's possible to look back at the conduct of any process and see who did what and when. As *Trial Manager*, Francesca has the necessary rights to act another role in the process – *Auditor* – and to browse the history so far. It's not long before she discovers that one of her investigators has been taking an inordinately long time taking part in a joint decision that sits on the critical path. Stepping out of that role and back into the *Trial Manager* role, she checks the background of the investigator to find that this is his first experience of this sort of trial. Time to act ...

Spreading best practice

The Arts and Media Faculty at Wellow University has recently taken the opportunity presented by some internal reorganization to redefine its processes. Using Riva, they prepare a RAD for each of their processes.

The Faculty team quickly put the processes into the BPMS and start to get immediate benefits – they know that all the work items covered in their process architecture will be managed and tracked through to completion. Not surprisingly, news of the improvements they've made soon gets around, and before long they find themselves leading a 'best practice team' tasked with rolling out their processes to other faculties.

The related group in the Fine Arts Faculty, not known for their love of new technology, is the first to raise its performance. There are slight differences between the two faculties, but it's such a simple matter to add Fine Arts administrative staff as role actors in the BPMS that they simply pick up the best practice processes, with a view to making on-the-fly modifications not long after roll-out.

The Fine Arts group are most impressed by the fact that they need only register in the BPMS to start adopting the processes by simply playing their roles in them. No software had to be written for them, they simply started using the system with their normal browser at their PCs. After two months using the Arts and Media processes, they are soon adapting them for their own environment.

Building one-off collaborative processes on-the-fly

Dealing with an emergency is not just about each individual agency doing its job. Time and again rehearsals have shown that close collaboration between agencies is essential in mounting a rapid response.

An urgent message arrives at the Emergency Response Centre: Barrack Hill caves and the houses built over them have collapsed. The Response Centre Manager, Mark, turns straight to the BPMS. The bones of the collaborative process have been drawn up as a Riva model over the last two years and are now waiting in the BPMS, ready to be brought into play to coordinate the emergency services.

No two emergencies are alike – the process has to be constructed in broad outline and details added as the emergency unfolds. Mark has used the process authoring part of the BPMS many times before in rehearsals, taking just minutes to build an appropriate collaborative process, and it's not long before the BPMS is running the collaboration of teams from the fire services, cave rescue, the major utilities, and the hospital crisis centre. Even automatic information feeds to the media are already built into the process, and can start as soon as the process is kicked into action. Lives and property are saved.

Customer-focused processes from functional silos

People in need of medical care are frequently treated by staff from many different disciplines, and all too frequently these work almost in isolation from each other, even though they share a common objective in treating the patient. This isolation is reflected in separate processes, separate roles and separate information resources.

Each of the resulting 'silos' of activity may well optimize the performance of the individual service, but that doesn't necessarily mean that the total service delivered to the patient is the best it can be.

As a pilot study in more effective focusing of all services on the patient, the authorities in Barsetshire form a cross-functional team from the different groups that need to coordinate their work to deliver a service package to victims of strokes. Speech therapy, occupational therapy, physiotherapy, social services, the medical team, wheelchair mechanics and many more, first use Riva to develop a process architecture for their individual areas, and then work together to place these in a larger architecture for integrating stroke management services. As a result, they are able to construct a single process, focused on the patient, which integrates the processes of each individual group with very few changes.

In stage 2, Barsetshire puts a BPMS in place to run the overall stroke management process, principally by coordinating information flows between the groups and notifying each group when its contribution is required. In stage 3, processes from the individual groups are added to the BPMS, and gradually an integrated process develops and spreads.

The stroke patient now feels they are the focus of the work of all the teams that collaborate to improve their quality of life.

Built-in measurement leads to process improvement

The administration team for Rode University's Engineering Faculty has adopted the processes developed at Wellow University and is successfully running the administration of students, courses, awards, and modules using those processes. When the Vice-Chancellor announces that five per cent of funds will shift from administration to teaching, yet more savings have to be found and the Faculty administration team turn to the BPMS in which they play their processes.

As part of the initial Riva-based process definition work they have done, each process for dealing with a different type of work item has been defined in terms of the actions of a set of roles and the collaboration between those roles. Those processes are now playing in the BPMS. Thanks to the way Riva has grouped their work into CPs and CMPs, it's easy to identify where to insert 'measurement probes'. The collection of data about the performance of the processes immediately becomes automatic.

Some extra analysis activities are added to two management roles and within days performance hot-spots become visible and point at opportunities for process improvement. Almost invariably these occur at boundaries with other, non-BPMS processes owned by other groups, where collaboration can break down all too easily and impact performance. With diagnostic data to hand, the Faculty's processes are extended into the other groups and more reliable collaboration is soon in place, with cost savings following.

Keeping processes under audited control

In his role as Manager of Standard Operating Procedures (SOPs) at pharmaceutical company BenePharm, Sanjay is a man with processes for looking after processes. The industry regulator takes a very keen interest in how things are done in their labs, requiring documented processes and moreover this is where Sanjay comes in – processes for controlling those definitions themselves, the SOPs.

His group is small and their processes are relatively few, but managing SOPs involves a great many people around the business: front-line staff must draft them, business managers must approve drafts and changes, changes must be evaluated by an SOP committee and finally, of course, they have to be published on the intranet for people to be able to use them. And Sanjay has to be able to demonstrate to the regulatory inspector that all of this is being done in accordance with their SOP management processes – the SOPs for managing SOPs.

This is one load that the BPMS has taken off Sanjay's mind: with his processes loaded into the BPMS they run just as required – no chasing or checking by him or his staff. They can concentrate on the content and usage of the SOPs. And because the BPMS keeps a full record of what happens, that mandatory audit trail he needs comes for free.

Tutor:	Wouldn't that all be nice?
Pupil:	It would indeed.

References

Auramäki, E., Hirschheim, R. and Lyytinen, K. (1992) Modelling offices through discourse analysis: the SAMPO approach. *Computer Journal*, **35**, 4, 342–352.

Burlton, R. T. (2001) *Business Process Management – Profiting from Process.* SAMS, Indianapolis.

Checkland, P. and Scholes, J. (1990) *Soft Systems Methodology in action.* Wiley, Chichester.

Greenspan, S. (1985) *Requirements modeling: a knowledge representation approach to software requirements definition.* Technical Report CSRG-155, Computer Systems Research Group, University of Toronto.

Holt, A. W., Ramsey, H. R. and Grimes, J. D. (1983) Coordination system technology as the basis for a programming environment. *Electrical Communication*, **57**, 4, 307–314.

ISO 9001:2000 *Quality Management Systems – Requirements.*

Jackson, M. (1983) *System Development.* Prentice-Hall, Englewood Cliffs, New Jersey.

Jacobson, I., Ericsson, M. and Jacobson, A. (1994) *The Object Advantage.* Addison-Wesley, Wokingham.

Milner, R. (1999) *Communicating and Mobile Systems: The Pi-Calculus.* Cambridge University Press, Cambridge.

Office of Government Commerce (2001) *Business Systems Development with SSADM.* UK.

Oncken, W. (1987) *Managing management time: who's got the monkey?* Prentice-Hall, Englewood Cliffs, New Jersey.

Ould, M. A. (1995) *Business Processes.* John Wiley, Chichester.

Ould, M. A. (1999), *Managing Software Quality and Business Risk.* John Wiley, Chichester.

Ould, M. A. and Birrell, N. (1985 and 1988) *A Practical Handbook for Software Development.* Cambridge University Press, Cambridge.

Ould, M. A. and Roberts, C. (1987) Defining formal models of the software development process. In Brereton P. (ed.), *Software Engineering Environments.* Ellis Horwood, Chichester.

Ould, M. A. and Unwin, C. (1986 and 1988) *Testing in Software Development.* Cambridge University Press, Cambridge.

Patching, D. (1990) *Practical Soft Systems Analysis.* Pitman Publishing, London.

Roberts, N., Andersen, D., Deal, R., Garet, M. and Shaffer, W. (1983) *Introduction to computer simulation: the system dynamics approach.* Addison-Wesley, Boston, MA.

Smith, H. and Fingar, P. (2002) *Business Process Management: The Third Wave.* Meghan-Kiffer Press, Tampa, Florida.

Smith, H. and Fingar, P. (2004) *Outoperate your competition using the BPMS.* Available at http://www.BPTrends.com.

Winograd, T. (1987) A language/action perspective on the design of cooperative work. *Human-Computer Interaction*, **3**, 1, 3–30.

Winograd, T. and Flores, F. (1987) *Understanding Computers and Cognition.* Addison-Wesley, Reading, MA.

Index

BCS Products and Services

Other products and services from the British Computer Society which might be of interest to you include:

Publishing

BCS publications, including books, magazines and peer-review journals, provide readers with informed content on business, management, legal and emerging technological issues, supporting the professional, academic and practical needs of the IT community. www.bcs.org/publications

BCS Professional Products and Services

The BCS promotes the use of the SFIA*plus* IT skills framework which forms the basis of a range of professional development products and services for both individual practitioners and employers. This includes BCS Skills*Manager* and BCS Career*Developer*. www.bcs.org/products

Qualifications

Information Systems Examination Board (ISEB) qualifications are the industry standard both here and abroad, and with over 100,000 practitioners now qualified, it's proof of their popularity. There's a huge range on offer covering all major areas of IT. ISEB qualifications are for forward looking individuals and companies who want to stay ahead – who are serious about driving business forward. www.iseb.org.uk

The BCS Professional Examination is examined to the academic level of a UK honours degree and is the essential qualification for a career in computing and IT. Whether you seek greater job recognition, promotion or a new career direction, you'll find this is internationally recognised, flexible and suited to the needs of the IT industry. www.bcs.org/exam

European Certification of IT Professionals (EUCIP) is aimed at IT Professionals and Practitioners wishing to gain professional certification and competency development. www.bcs.org/eucip

European Computer Driving Licence (™) **(ECDL)** is the internationally recognised computer skills qualification which enables people to demonstrate their competence in computer skills. ECDL is managed in the UK by the BCS. ECDL Advanced has been introduced to take computer skills certification to the next level and teaches extensive knowledge of particular computing tools. www.ecdl.co.uk

Networking and Events

The BCS's national network of branches and specialist groups enables members to exchange ideas and keep abreast of latest developments. www.bcs.org/sg

The Society's programme of social events, lectures, awards schemes, and competitions provide more opportunities to network. www.bcs.org/events

Further Information

This information was correct at the time of publication but could change in the future. For the latest information, please contact: **The British Computer Society, 1 Sanford Street, Swindon, Wiltshire, SN1 1HJ**, **U.K.**

Telephone: +44 (0)1793 417 424
Email: bcs@hq.bcs.org.uk
Web: www.bcs.org

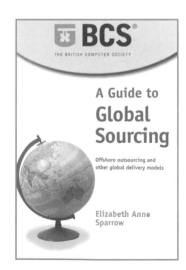

A Manager's Guide to IT Law

- IT Contracts
- Instructing an IT Consultant
- Intellectual Property Law for Computer Users
- Setting Up Joint Ventures
- Resolving Disputes
- Implementing New Systems
- Avoiding Employment Problems
- Systems Procurement Contracts
- Escrow
- Outsourcing
- Data Protection

Edited by Jeremy Newton & Jeremy Holt
Cover price £25. ISBN: 1-902505-55-7
orders@yps-publishing.co.uk

···▶ **www.bcs.org/books/itlaw**

BCS
THE BRITISH COMPUTER SOCIETY

BCS

It's about
setting standards
Not standing still

There has never been a more rewarding time to join the BCS.
We've changed our constitution to become more dynamic.
We've streamlined our membership structure to make the benefits of joining more
professionally rewarding.

Our position as the industry body for information technology professionals, and the leading
chartered engineering institution for IT, has never been stronger. Join today and stand out
as a standard-setting professional.

You can find out more, or join straightaway, simply by visiting our website at
www.bcs.org/gofurther, or by emailing **gofurther@hq.bcs.org.uk**
or calling free on **0800 056 4322**

 going further **together**